Anaerobic Digestion - Biotechnology for Reactor Performance and Environmental Sustainability

Edited by Sevcan Aydın

Published in London, United Kingdom

Anaerobic Digestion - Biotechnology for Reactor Performance and Environmental Sustainability
http://dx.doi.org/10.5772/intechopen.111094
Edited by Sevcan Aydın

Contributors
Moses Jeremiah Barasa Kabeyi, Oludolapo Akanni Olanrewaju, Yang Min, Vhutshilo Nekhubvi, Varinder Singh Saimbhi, Sevcan Aydin, Şuheda Reisoglu, Vadym Poliakov

First published in London, United Kingdom, 2024 by IntechOpen
IntechOpen is the global imprint of INTECHOPEN LIMITED, registered in England and Wales, registration number: 11086078, 5 Princes Gate Court, London, SW7 2QJ, United Kingdom

British Library Cataloguing-in-Publication Data
A catalogue record for this book is available from the British Library

Additional hard and PDF copies can be obtained from orders@intechopen.com

Anaerobic Digestion - Biotechnology for Reactor Performance and Environmental Sustainability
Edited by Sevcan Aydın
p. cm.
Print ISBN 978-1-83769-231-6
Online ISBN 978-1-83769-232-3
eBook (PDF) ISBN 978-1-83769-233-0

For EU product safety concerns:
IN TECH d.o.o., Prolaz Marije Krucifikse Kozulić 3, 51000 Rijeka, Croatia,
info@intechopen.com or visit our website at intechopen.com.

Meet the editor

Sevcan Aydin is a professor in the Biology Department, Istanbul University, Turkey. She earned her bachelor's degree from the Biology Department, Ege University, Turkey, followed by a Ph.D. from the Environmental Biotechnology Program, Istanbul Technical University in 2015. Dr. Aydin achieved a professorship in environmental engineering, and her research interests encompass environmental biotechnology, renewable energy production, bioremediation, and environmental engineering. She has published forty-nine scientific articles in high-impact journals and supervised two doctoral and eight master's theses. Dr. Aydin teaches biotechnological approaches to water treatment, environmental technologies, and bioenergy production methods. Her specialization lies in enhancing the bioenergy potential of different wastes and developing cost-effective methods for the bioremediation of polluted soil.

Contents

Preface

Anaerobic digestion is a process that converts organic matter into biogas, a valuable source of energy that contains mainly methane and other impurities such as hydrogen sulfide, carbon dioxide, water vapor, siloxanes, hydrocarbons, oxygen, ammonia, carbon monoxide, and nitrogen. Anaerobic digestion is not only a natural and eco-friendly way of producing biogas but also a way of treating waste materials and reducing greenhouse gas emissions, which are major challenges for the world in the era of fossil fuel depletion, climate change, and environmental pollution. Moreover, anaerobic digestion has many other benefits, such as reducing the volume and mass of waste, eliminating odors and pathogens, generating digestate as a nutrient-rich fertilizer, conserving land and water resources, creating jobs and income opportunities, and mitigating climate change. Therefore, anaerobic digestion is a promising technology for the future of waste management and energy production. As the demand for biomass-based energy has increased in recent years, due to the challenges posed by the depletion of fossil fuels, climate change, and environmental pollution, this book, *Anaerobic Digestion – Biotechnology for Reactor Performance and Environmental Sustainability*, offers a comprehensive overview of the latest developments and approaches in anaerobic digestion. The book consists of six chapters, each covering a different aspect of anaerobic digestion technology.

Chapter 1 examines the most prevalent physicochemical technologies for biogas upgrading and CO2 removal, along with some novel findings and possible improvements. It also evaluates different methods of producing biomethane from biogas. It gives a thorough description of the main principles of different biogas upgrading approaches, and their biomethanation efficiency. It also addresses the challenges and opportunities for further development of these technologies.

Chapter 2 focuses on the sludge-water mixing process in anaerobic reactors, which is essential for the anaerobic degradation of organic matter in wastewater. It explains the concept of CSTBR (completely stirred tank bioreactor) mode, which requires low mixing intensity and good sludge morphology and mass transfer. It compares two types of confined sludge anaerobic reactors, up-flow anaerobic sludge blanket (UASB) and internal reflux packed-bed, which can achieve CSTBR mode by organizing the water flow with anaerobic sludge in attached or aggregated state.

Chapter 3 presents the results of a research study and explores the chemical and thermal properties of cow manure–water mixtures for biogas production. The chapter shows how the specific heat capacity and chemical composition of cow manure–water mixtures vary with temperature and mixing ratio, and how they affect the biogas yield and quality.

Chapter 4 provides mathematical modeling and applied calculation of bioconveyer and anaerobic biofiltration. The chapter develops and presents a theoretical framework along with an engineering methodology for evaluating the performance of anaerobic biofiltration. The findings contribute to the foundation for the applied optimization of bioconveyer plant parameters, providing essential insights to enhance their operational efficiency.

Chapter 5 discusses biomass gasification, a process that converts organic matter into a combustible gas, which can be used for various applications such as cooking, lighting, heating, and power generation. It is a sustainable and environmentally friendly technology that can enhance rural livelihood security by providing energy and income opportunities. This chapter explores the different types of biomass gasification, namely biochemical and thermochemical, and their advantages and challenges. It also presents an assessment of the impact of biomass gasification on the economic, social, and environmental aspects of rural households, based on a survey conducted in India. The chapter shows how biomass gasification can improve agricultural productivity, reduce fuel consumption and emissions, and increase the income and savings of the rural population.

Chapter 6 reviews the role and applications of bacteriophage in microbial community dynamics, a type of virus that infects and kills specific bacteria, in improving the anaerobic digestion process. The chapter discusses the challenges and opportunities of using bacteriophage as biotechnological tools to enhance biogas production and quality, control pathogens and biofilm, and manage imbalances in anaerobic reactors.

This book is for researchers, students, and practitioners interested in learning more about anaerobic digestion technology and its applications. It provides a comprehensive and up-to-date source of information and knowledge on the biotechnology of anaerobic digestion, with the aim to inspire further research and innovation in this field. We hope that readers will find this book useful and informative and that it will contribute to the advancement and dissemination of anaerobic digestion technology for the benefit of society and the environment.

Dr. Sevcan Aydin
Professor,
Division of Biotechnology,
Biology Department,
Istanbul University,
Istanbul, Turkey

Chapter 1

Biomethane Production and Applications

Moses Jeremiah Barasa Kabeyi
and Oludolapo Akanni Olanrewaju

Abstract

Biomethane production generally involves the cleaning to remove minor unwanted components of biogases such as hydrogen sulfide (H_2S) and moisture (H_2O) and upgrading in a process that involves the removal of carbon dioxide (CO_2) to increase the concentration of CH_4 to 95–99% and reduce CO_2 concentration to 1–5%, with little or no hydrogen sulfide (H_2S). Biomethane gas is a flexible and easy to store fuel having similar properties and applications as natural gas with no need to modify the settings for natural gas devices and equipment. Biomethane can be used for industrial and domestic applications ranging from thermal and power generation and feedstock for processes like the Fischer-Tropsch (FT) for fuel manufacturer and direct power generation in hydrogen or biogas fuel cells like production of green hydrogen. Therefore, biomethane promises to play a leading role in the energy transition through hydrogen, electricity, and other renewable fuels production. Biomethane production by biogas upgrading methods include the pressure swing adsorption, which has an option of temperature swing adsorption, absorption technics based on amine, membrane separation, cryogenic separation, and biological separation. The technology adopted may depend on factors such as costs, quality of products, location, and technology maturity and requirements.

Keywords: biogas, biogas cleaning, biogas purification, biomethane, anaerobic digestion, biogas enrichment

1. Introduction

The production of biogas has been growing and so is the demand for upgraded biogas for applications like vehicle fuel or injection to the natural gas grid. Biogas has to be upgraded to facilitate efficient use in these applications by removal of carbon dioxide which is inert yet it constitutes a significant portion of raw biogas at the expense of methane [1, 2]. Biogas is a mixture of gases produced by action of microorganisms through anaerobic digestion which is a complex process made up of four stages i.e.: hydrolysis, acidogenesis, acetogenesis and methanogenesis leading to biogas. The composition of biogas is influenced by the type of feedstock used and anaerobic digestion process control [3]. Other than production in anaerobic digesters, biogas can also be produced from landfills and through biomass thermal pyrolysis and gasification processes. The mixture produced generally consists of 30–75% methane (CH_4), 25–55% carbon dioxide (CO_2) and other constituents or impurities like hydrogen (H_2), oxygen (O_2), nitrogen (N_2), hydrogen sulfide (H_2S), water (H_2O) and ammonia (NH_3), dust particles, siloxanes, aromatic and halogenated compounds, which are often in tiny quantities [4–6].

Anaerobic digestion is a sustainable process used for simultaneous treatment and production of biogas energy resource. Biogas is a renewable energy resource produced by anaerobic digestion and has methane as the main component with impurities like hydrogen sulfide, carbon dioxide water vapor, siloxanes, hydrocarbons, oxygen, ammonia, carbon monoxide and nitrogen. The energy content of biogas is reduced by the impurities while others cause operational challenges to combustion systems like corrosion [7, 8]. This makes it important to apply different technologies to remove these harmful and undesirable impurities [9]. There are many established and developing physicochemical technologies for biogas operation and maintenance costs, energy requirements, efficiency of removal and other parameters [9, 10].

Biomethane or upgraded biogas is made by qualitative processing of raw biogas through several steps to remove impurities mainly carbon dioxide. Biomethane as a fuel provides new opportunities at different levels for the society. In this chapter, various methods of biomethane production are described and compared [11, 12]. There is growing interest in the use of biomethane as a renewable substitute of natural gas in applications like transport fuel which has created demand for biogas upgrading. This chapter is a critical review that summarizes the state-of-the-art technologies used in cleaning and upgrading. Covered in the review are the description of biomethane production methodologies, scientific and technical outcomes related to them, bio-methanation efficiency, challenges and feasibility of the technologies [8, 13].

Biogas is a renewable energy carrier that can be exploited directly as a fuel or as a feedstock for production of hydrogen or synthesis gas. The main constituents of biogas are carbon dioxide (CO_2) and Methane (CH4), but there are quantities contaminants like such as hydrogen sulfide (H_2S), ammonia (NH_3), moisture and siloxanes whose existence and composition is a function of the source of biogas like landfills, anaerobic fermentation of manure [14, 15]. The contaminants have the following undesirable effects to biogas applications a fuel;

i. They can be detrimental to any biogas thermal or thermo-catalytic conversion device (e.g., corrosion, erosion, fouling); and

ii. Generation harmful environmental emissions. It is therefore important to include biogas purification steps upstream of its final use processes [9].

Other than methane which is the main energy source in biogas, raw biogas has impurities that are noncombustible while others are harmful to the equipment and environment and should therefore be removed to make it suitable for wide range of applications in heat and power generation [8]. The treatment and purification/upgrading pathways and applications are summarized in **Figure 1** below.

From **Figure 1**, it is shown that biogas treatment mainly involves desulphurization and drying of raw biogas. Making it an ideal feedstock for applications like boiler fuel, cogeneration (CHP) and biogas reforming for production of hydrogen and other fuels and as s fuel for direct combustion processes like boilers for heat and power production.

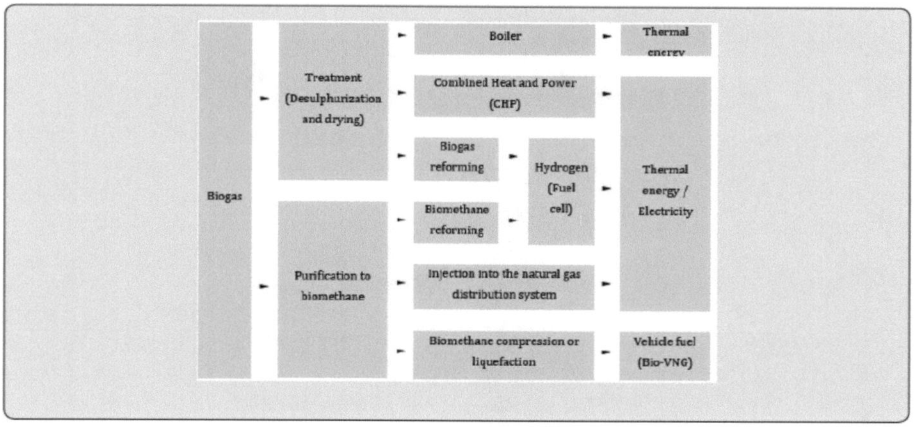

Figure 1.
Main biogas potential applications.

Purification on the other hand leads to the production of biomethane which is an ideal feedstock for biomethane reforming for hydrogen, production of compressed liquid bio-natural gas and liquefied bio natural gas as well a direct substitute for natural gas fuel applications including injection to natural gas pipelines [2, 6, 16].

2. Technologies for biogas cleaning, drying and purification

Controlling the level of impurities in biogas is essential for success of its recovery. The implementation of treatment and purification technologies must consider the requirements of each specific application of biogas. These technologies aim to adjust the calorific value and remove contaminants that affect the quality of biogas and the useful life of the equipment. The most demanding techniques aim to purify biogas to obtain biomethane. Currently, different techniques that allow the treatment and purification of biogas are commercially available [12, 17, 18].

There are many technologies available at commercial and laboratory scale for the treatment and purification of biogas. These methods include condensation, absorption and adsorption processes for raw biogas. Hydrogen sulfide (H_2S) removal can be done by biological in situ, ex situ, biofilter and biological gas scrubbing techniques. Purification which involves mainly removal of CO_2 can be done by absorption i.e. amine and pressurized water scrubbing and amine scrubbing, membrane permeation and cryogenic method [2, 19–21]. **Figure 2** summarizes the various methods for treatment/cleaning and purification of raw biogas.

From **Figure 2**, it is noted that treatment or cleaning mainly involves removal of moisture and H_2S while upgrading or purifications mainly targets the CO_2 for removal. The choice of biogas treatment and purification technology is a function of factors like the amount of biogas produced, its composition, the level of purification required, and process costs in terms of capital, energy consumption and operational expenditure (CAPEX and OPEX). In biomethane requirements, a combination of processes is used, as no technology can remove all contaminants from biogas [1, 2].

There are simpler and cheaper technologies available for treatment of biogas with the objective of cleaning biogas for sensitive applications [2]. The degree of biogas

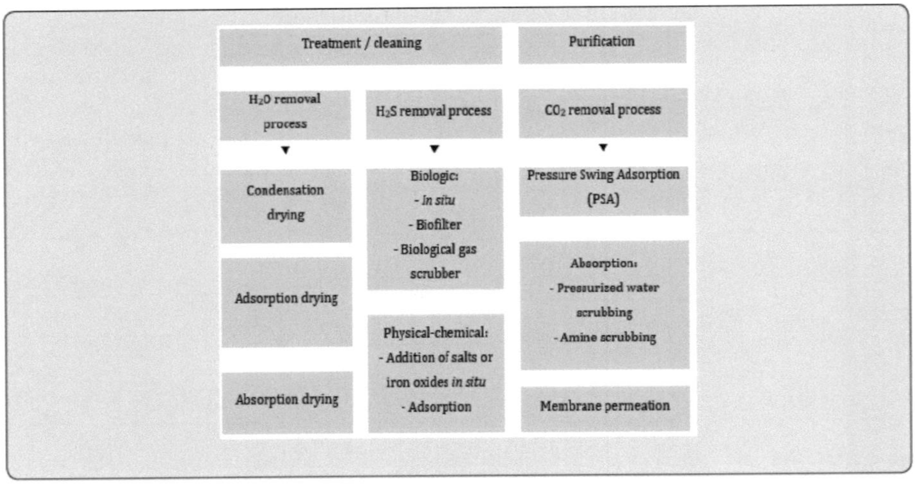

Figure 2.
Main technologies to biogas treatment and purification.

treatment depends on the intended application and the initial composition of raw biogas. Most common treatment involves the removal of H_2O and H_2S [4, 12].

2.1 Moisture (H_2O) removal

Biogas may contain moisture concentration of between 3 to 10%. Water removal is usually carried out at an early stage of the treatment, to protect the downstream equipment against corrosion and allow the biogas feedstock to fulfill the requirements of subsequent purification steps. **Table 1** is a summary of advantages and disadvantages of processes for H_2O removal from biogas.

From **Table 1**, the strengths and weaknesses of moisture removal techniques from biogas are presented. Moisture can be removed from raw biogas by condensation, adsorption and absorption. Condensation has high energy consumption and is expensive in terms of investment and maintenance, but the process is simple and effectively removes hydrocarbons and oil particles. The adsorption process has low operating costs, has high removal rate and the adsorbent materials are regenerated as main advantages, although it has high investment costs and requires the prior removal of oil. The Absorption process of moisture removal from raw biogas effectively eliminates hydrocarbons, and like the adsorption method has high removal rates and absorption materials can be regenerated. However, the method has high investment costs, making it only viable at high biogas flow rates. The regeneration of adsorption materials is done at high pressure and temperature making operations and maintenance expensive [7, 22].

i. Condensation

In condensation, separation of steam and water from biogas is affected through by use of cyclone separators. Condensation of water can be improved further by cooling biogas below the dew point of the gas. For this purpose, cooling pipes are installed with a slope and a purging system to collect the condensate collected [1, 6].

Process	Advantages	Disadvantages
Condensation drying	• Simple process, suitable for any biogas flow; • Elimination of hydrocarbons and oils particles; • Application as pre-treatment in all systems.	• High energy consumption; • Requires installation of long tubes with slope and freeze-resistant; • High investment and maintenance costs.
Adsorption drying	• Adsorbent materials can be regenerated; • High removal rate, which allows the process to be applied to any type of biogas use; • Low operating cost.	• Requires prior removal of particles and oil; • High investment cost; • Suitable for small or medium biogas flows.
Absorption drying	• Materials can be regenerated; • High removal rate, which allows the process to be applied to any type of biogas use; • Elimination of hydrocarbons particles.	• High investment cost; • Economic viability only for high biogas flow rates; • Absorbent material regeneration carried out at high pressure and temperature.

Table 1.
Advantages and disadvantages of the main H_2O removal processes from biogas.

ii. Adsorption

Cylindrical reactors containing adsorbent materials are used in adsorption drying process. Commonly used adsorption materials are silica gel, activated carbon, aluminum oxides, magnesium oxides and zeolites. The adsorption materials are installed in a fixed bed, that can be exchanged and regenerated when it gets saturated. The system can also operate alternately with two columns, where one has the adsorption material at room temperature and pressure between 6 bars and 10 bars, and the other column is a standby unit where regeneration is done [1, 23].

iii. Absorption

In absorption drying biogas flows through an absorption tower, in countercurrent with a solution of glycol or other hygroscopic materials. In the process moisture or steam and hydrocarbons are chemically absorbed. This method was originally used to dry natural gas. Absorption operations take place at high pressure of between 20 and 40 bars, while regeneration occurs at around 200°C [1, 2].

2.2 H_2S removal process

Hydrogen sulfide is (H_2S) is a gaseous chemical found in many fuel gases, biogas, natural gas, syngas, coke oven gas, landfill gas, refinery gas, and wastewater steams among others etc. [24]. Hydrogen sulfide is flammable, toxic, and extremely hazardous and should therefore be captured and removed from biogas. The challenge and need to H_2S has led to the development of different materials and methods over the years for its removal. Some alkanolamines are used as absorbents and while metal oxides are used as adsorbents [17, 18, 25]. The removal of H_2S from fuels is imperative in terms of both safety and economics. The main challenge of H_2S in application of biogas is its inherent tendency to form an acidic solution in the presence of water which leads to pipelines and equipment corrosion [8, 25].

H_2S removing process is divided into two broad levels or categories on the basis of intended application for produced biogas. Hydrogen sulfide (H_2S) removal is normally done by means of biological or physical- chemical processes, and can be classified as external or internal depending on whether it is done outside or inside the anaerobic bio-digester. The first level involves production of biogas with, H_2S concentration of below 500 ppm, and can reach as low as 100 ppm. The second level involves reduction in H_2S concentrations less than 0.005 ppm, which are typical specifications and requirements for biomethane gas [4, 24].

2.2.1 Biological removal of H_2S

There are various established methods for biological removal of H_2S from raw biogas. They include in situ, biofilter, biological scrubber, and ex-situ techniques.

i. In Situ methane enrichment

Situ is a biotechnology based on the direct injection of pure air or oxygen and in the process, the bacteria that oxidize H_2S develop with the presence oxygen, leading to the biological removal process of H_2S, to produce sulfur (S) which leaves the digester via the digested. The microorganisms are widely found in the anaerobic environment present in bio-digesters [4]. In situ desorption technology is yet to be fully developed even though it has been around for over 20 years. In-situ is based on the greater solubility of CO_2 over CH_4 in water. The process set up includes an anaerobic digester linked or connected to an external desorption unit. Sludge transported to an aerated desorption column from the digester. Nitrogen or air flowing in counter-current mode and dissolves the CO_2 from the sludge in the desorption unit. The sludge desorbed sludge is pumped back into the digester to reabsorb amore CO_2, and the sludge as the sludge is continuously recycled in the desorption column. It is possible to strip out H_2S with dissolved CH_4 and CO_2 from the recirculating sludge by applying large quantities of air or N_2, causing reduction in the H_2S and CO_2 concentration [26, 27].

Therefore, in Situ is a biotechnology which works by direct injection of pure air or oxygen and causing bacteria to oxidize H_2S leading to biological removal H_2S, to produce sulfur (S) which exits the digester via the digested. These microorganisms are widely present in anaerobic environment in bio-digesters [4].

ii. Biofilter

In the bio filter technology biogas is passed through a column having a synthetic material, in the form of a biofilm. The parallel or countercurrent flow system is used to maintain the humidity and nutrients, that are essential for the microorganisms that degrade of H_2S [4]. The purification system consists of a bioreactor where sulfur-oxidizing bacteria like the Thiobacillus, Pseudomonas and Acidothiobacillus are immobilized on a carrier. In the process, moisturized biogas is injected from the bottom of the bio filter and forced through a moist, packed bed with microbial biofilm which purifies biogas. The bed material is used to supply nutrients or nutrient solution added from the top occasionally. Oxygen whose concentration is 5–10% of volume is supplied by injecting air directly into the gas stream [3].

iii. Biological gas scrubber

In biological gas scrubber, a two-stage system is used to remove H_2S. In the first stage H_2S scrubbing column, applies sodium hydroxide solution while activated sludge is used in the second stage which is injected with injected with air, because the microorganisms used are aerobic, leading to the solution regeneration [4]. Bio scrubber system is applied in the removal of compounds like ammonia, amines, hydrocarbons, hydrogen sulfide and odorous contaminants. Bio scrubbing system consists of two reactor units with the first reactor as absorption tower where the pollutants are absorbed in a liquid phase before it goes to the second reactor which is activated sludge. Degradation occurs in the activated sludge reactor where microorganisms like Thiobacillus and Thioalkalivibrio) grow in suspended flocks. The effluent generated is recirculated back to the absorption tower. In the removal of H_2S, a sedimentation tank is installed after the second reactor for collection of elemental Sulfur with O_2 being used as the oxidant. Optimal microbial growth and activity are maintained addition of oxygen, nutrients and pH regulation together with continual purging of by-products and excess biomass out of the system [3].

iv. Ex-situ

Ex- situ biogas cleaning and upgradation relies on supply of carbon dioxide from external sources and hydrogen in an anaerobic reactor, which eventually contributes to their conversion to methane. The ability of ex situ process to manage high concentrations of influent gases, reduces retention time to about 1 hour leading to a smaller device for upgrading. Depending mainly on the reactor used, the ex-situ technology can produce methane with final purity of 79–98%, the main challenge facing this technology is low gas–liquid mass transfer rate [26]. Therefore, ex-situ is more of an upgrading than cleaning method although it can do both by design. The advantages and disadvantages of the biological H_2S removal processes are summarized in **Table 2** below.

From **Table 2**, the three discussed biological methods for H_2S removal have significant differences in terms of use of chemicals, operation and maintenance costs and product quality. Biofilter and biological gas scrubber techniques need external oxygen injection. The main advantage of in-situ method is that it has low investments and maintenance costs and does not require chemicals.

2.2.2 Physical-chemical removal of H_2S

These techniques involve use of salts or iron oxides or sulfide precipitation is used to remove H_2S inside the digester. Iron oxides or salts are added that react with H_2S, to produce non-soluble compounds, like iron sulfides, that precipitate and are removed

Process	Advantages	Disadvantages
In situ	• A simple process • Low investment and maintenance cost • Does not need chemicals	• Oxygen injection my affect the anaerobic digestion and can oxidize methane • Potentially explosive mixtures may occur • Cannot achieve biomethane purification level
Biofilter	• Enables the removal of ammonia • Does not require chemicals • The injection of oxygen is external to the digester hence no negative effect to the digestion	• Requires nutrients renewal hence more operation and maintenance cost • Only suitable for small biogas flows • Injection of air at high levels through the biofilter is not suitable for biomethane production
Biological gas scrubber	• Oxygen introduced is external to the process and has no negative impact to the digestion • Good for high biogas flow rates • The process can attain purity requirement for biomethane	• Uses chemicals • High operations and maintenance costs • The process needs fresh water introduction
Ex citu	• Can be used to attain high methane purity levels needed for biomethane • Requires smaller devices due to lower retention time	• The process relies on carbon dioxide supplied from external sources hence an extra cost • The process has low gas–liquid mass transfer rate

Table 2.
Biological process for removal of H_2S (summary by the author).

together with effluents from the biodigester. Through direct dosing, chemicals are added to a reactor installed in the biogas line. The H_2S adsorption process is achieved by retention in a solid form having a large surface area or in materials with high internal porosity. Activated carbon and iron oxides are the common adsorbent materials applied in the process. Activated carbon enable production of low concentrations of H_2S based on the catalytic oxidation of H_2S on the surface of the activated carbon, which is easy to impregnate with catalysts that speed up the reaction and improve the process capacity [4]. The advantages and disadvantages of each physical- chemical H_2S removal process is summarized in **Table 3**.

From **Table 3**, it is noted that there are broadly two main approaches of physical-chemical methods of H_2S i.e. addition of salts or iron oxides in situ and adsorption. The addition of chemicals does not be used to attain biomethane quality although the process is simple and cheap. The adsorption method is moderate in cost, attains high removal rate for H_2S to attain biomethane level of purity but incurs high energy costs and high operation and maintenance cost related to replacement of the adsorbent.

3. Biogas upgrading methods

Production of biogas and use has several environmental, social and economic benefits. It is a source of renewable energy, and its production is also considered as a manure production factory. Biomethane has wider industrial applications hence biogas up-gradation is desirable [2]. The main drivers of biogas up-gradation is rapid increment in the price of fossil fuels and growing concerns over global climate change due to greenhouse gas emissions. Biomethane has opened a new window for the replacement of natural gas from the energy mix. There are multiple biogas

Process	Advantages	Disadvantages
Addition of salts or Iron oxides in situ	✓ The process is simple and cheaper ✓ No need for oxygen injection ✓ Low maintenance costs	✓ Cannot attain biomethane purity levels ✓ Forms precipitates within the digestor and hence handling issues ✓ The process uses chemicals
Adsorption	✓ Has moderate investment costs requirements ✓ Has got high rate of removal ✓ Can attain biomethane quality and standards ✓ Oxygen injection does not affect the use of doped activated carbon	✓ The process has high energy consumption ✓ Extra cost incurred to renew absorbent in form of operation and maintenance ✓ Extracted sulfur cannot be used

Table 3.
Advantages and disadvantages of physical-chemical methods for H_2S removal (summary by the author).

up-gradation technologies which are available on commercial scale while others are still developing and are at laboratory scale. The technologies that are widely accepted have attained prominence on their operational efficiency and reliability, merits and demerits and future outlook [10, 18, 21].

Biomethane production requires more complex and expensive techniques compared to biogas treatment methods aimed at attaining high degree of purity for biogas. Biogas upgrading combines biogas treatment and purification processes to remove other gases from biogas, thence separate methane (CH_4) and effectively increasing its heating value. Purification of biogas involves removal of carbon dioxide which is mandatory for biomethane to substitute natural gas in pipeline system for natural gas distribution and use as a fuel for applications like vehicle fuel [28].

3.1 Chemical scrubbing

Chemical scrubbing systems use aqueous organic or inorganic compounds to bind the CO_2 or H_2S molecules existing in biogas. The most commonly used scrubbing systems use organic compounds, namely aqueous amine solutions like diethanolamine (DEA), monoethanolamine (MEA) or methyl diethanolamine (MDEA). Most scrubbing systems using amine solutions have an absorber unit maintained between biogas pressure of 1–2 bars which is injected to the tank bottom with amine solution flowing from the top and a stripper. The solute and solvent (CO_2) undergo a reversible exothermic chemical reaction with the product amine solution which is rich in CO_2 and H_2S which proceeds to the stripper for regeneration operating at a pressure of 1.5–3 bars and temperature of 120-160°C [6].

The heat in the stripper disrupts the chemical bonds formed in absorption phase and which creates steam having CO_2. Upon cooling this steam, the CO_2 is released while the condensate is recirculated back to the stripping column. Some commercial systems can cope with biogas with H_2S content of up to 300 ppm/v, but H_2S poisons the amine, cause corrosion and increase the system energy requirements hence the need to remove H_2S before amine scrubbing [3, 25].

The advantages of the chemical scrubbing using amine solutions include high selectivity of the amines by CO_2 and the substantial extraction compared to other methods e.g. two times more CO_2 per unit volume is absorbed compared with water.

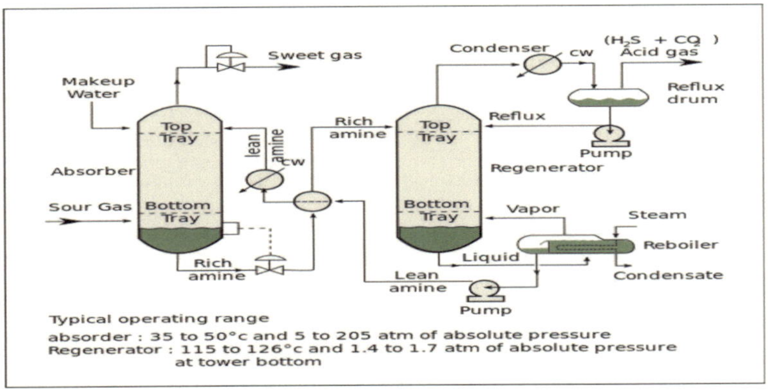

Figure 3.
Chemical scrubbing with amine solution [29].

The process has significantly low energy requirements mainly because of exothermic reactions and low process pressure operations i.e. 1–2 bar for absorption column and 1.5–3 bar in the stripping column. The draw backs include high energy requirements for solvent regeneration, expensive amine solvents and losses of solvents to evaporation which increase operation costs [3]. **Figure 3** shows the process of chemical scrubbing using amine solution.

From **Figure 3**, it is noted that the main system elements for the chemical scrubbing with amine solution are the adsorption column, heater, a cooler, a stripping column, and a heating medium for the stripping column which may be hot water, oil or steam.

In chemical scrubbing (CSC), there is reversible reactions between absorbed substances and solvent used. The commonly used biogas upgrading absorption solutions is based on amines i.e. methyl diethanolamine diethanolamine, monoethanolamine, and piperazine. For amine scrubber an absorber tank is used in which carbon dioxide is absorbed from the biogas operating at 20–65°C and 1–2 bar, then followed by a stripper where carbon dioxide is released by heating the stream. Chemical scrubbing with amine facilitates production of high concentration of methane concentration in biomethane greater than CH4 > 99%. The limitation of chemical scrubbing needs pre-treatment stage, to remove H_2S and has got high operational and investment costs [30].

The process is similar to pressurized water scrubbing, but is a chemical absorption technique. The solution absorbs CO_2 in biogas, by chemical reaction between amine and CO_2. The absorber is maintained at operating pressure 1–2 bar while the stripper maintained 1.5–3 bar. The process is exothermic, causing temperature rise of amine solution and higher efficiency since the reaction between amine and CO_2 increases with increase in temperature. Hydrogen sulfide (H_2S) should be removed prior to the reaction to avoid poisoning the amine solution [1, 7].

3.2 Organic physical scrubbing

Organic physical scrubbing work on the same principle with water scrubbing with the difference being the use of an organic solvent with higher affinity for H_2S and

CO_2. Methanol and dimethyl ethers of polyethene glycol (DMPEG) mixtures are all used in biogas upgrading. The process simultaneously absorbs hydrogen sulfide, carbon dioxide, and water due to their higher solubility in polyethene glycol than methane. Examples of commercially available organic physical scrubbing products Selexol® and Genosorb®. These products exhibit high hi solubility of NH_3 and CO_2 compared to H_2O. Selexol® can absorb three times more CO_2 than water hence lower liquid requirements which requires a smaller upgrading [3, 25].

The challenge associated with high solubility of carbon dioxide in organic solvents is difficulty to regenerate organic solvents. Higher solubility of H_2S compared to CO_2 in Selexol® leads to increased separation temperatures during the regeneration of the solvent hence higher energy consumption. It is therefore advice able to remove H_2S before the gas is treated with the solvent. The Selexol process may also be configured to remove H_2S selectively, or non-selectively in order to remove both CO_2 and (H_2S) [25].

In the first stage, raw biogas is compression and cooled to (7–8 bar, 20°C), before injection to the bottom of the absorption column. Since temperature affects Henry's constant the organic solvent is cooled down before it is fed to the column. The desorption column is used to regenerate the organic solvent by heating it to 80°C and reducing pressure to 1 bar. This leads to final methane content of 96–98.5% and less than 2% CH_4 losses, in an optimized full-scale plant [3]. **Figure 4** shows that the organic physical scrubbing method.

From **Figure 4**, it is noted that the main elements of the organic scrubbing method are Sulfur absorber, CO_2 absorber, H_2S concentrator, H_2S stripper, stripper reboiler reflux pump and reflux accumulator. In organic physical scrubbing, CO_2 in raw biogas is absorbed in an organic solvent e.g. a mix of dimethyl ethers of polyethylene glycol [29].

Concerns over the environment has motivated a shift from the use of conventional solvents to green solvents. This includes the use of deep eutectic solvents (DESs), consisting of two or more components, which are mainly hydrogen bond donors and acceptors [9, 25]. It desirable for the solvents to have a lower melting point, very low

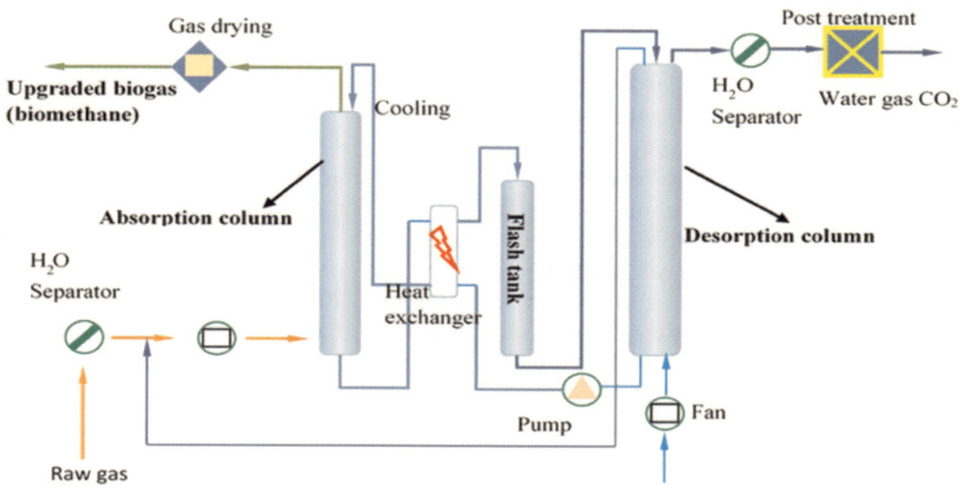

Figure 4.
Organic physical scrubbing [1].

vapor pressure and preferably be biodegradable. It is through selection of best fit hydrogen bond donors and hydrogen bond acceptors and donors, that the DESs can be appropriately engineered to yield desired thermodynamic and physical characteristics. It is also possible to remove other biogas contaminants by appropriate process modifications [3, 9].

3.3 Pressure swing adsorption (PSA) and vacuum swing adsorption (VSA)

The principle upon which the pressure swing adsorption (PSA) is the adsorption of CO_2 compared to CH_4 under conditions of high pressure due to differences in molecular characteristics and the affinity of the adsorbent material used. The vacuum swing adsorption (VSA) is based on the same principle is except that it operates under vacuum during the desorption step. Materials used as adsorbent matter PSA is required to have high surface area, e.g. alumina, silica gel, activated carbon, zeolite, polymeric sorbents and carbon molecular sieves [10, 31].

Pressure swing adsorption is a technique that works by selective adhesion of one or more components of the mixture, on the surface of a micro-porous solid. In this case the material for biogas upgrading is typically equilibrium-base adsorbents. The adsorbent pores allow an easy penetration of the carbon dioxide molecules but filters the larger methane molecules. The molecular sieve materials used include zeolites and activated carbon which act as the adsorptive materials for biogas upgrading. The process requires a pretreatment step because the materials used in PSA plants foul in the presence of raw biogas impurities. The pressure swing technology can achieve 95–99% methane purity for upgraded biogas which meats the typical technical specifications for the grid injection [25]. The main limitations of the pressure swing process are the pre-treatment requirement and extensive process control making the process expensive. To reduce operational costs, the temperature swing adsorption (TSA) is used instead. Temperature swing adsorption works at constant pressure and needs thermal energy to regenerate the adsorbent material making it suitable in applications having cheap heat source [30, 31]. The pressure swing method is illustrated in **Figure 5**.

From **Figure 5**, it is noted that the main processes and elements of the compressor for raw biogas, chambers for absorption, depressurization, desorption, pressurization, and vacuum pump for extraction of vent gases [10]. The characteristics of PSA unit include feeding pressure, cycle time, purging pressure, adsorbent, and column interconnectedness among other things [29].

Pressure swing adsorption (PSA) is a dry method used to separate gases on basis of their properties. Raw biogas is compressed to an elevated pressure and supplied to an adsorption column that retains CO_2 but leaves CH_4. Once the column material is saturated with CO_2, pressure is released hence CO_2 is desorbed and fed to the off-gas stream. Multiple columns can be applied columns are needed for continuous operation allowing them to be closed and opened consecutively [29].

The Pressure Swing Adsorption (PSA) technology makes use of the ability of porous adsorbent medium to adsorb specific molecules out of raw biogas and then release through the application of different pressure levels. For the case of raw biogas upgrading [1, 29]. For the biogas upgrading process, the operation is based on the different molecular dimensions of CO_2 which is 0.34 nm, methane CH_4 with 0.38 nm. The application of adsorbent material with cavities of 0.37 nm facilitate retention of CO_2 in the pores, as methane flows out with no retention. The most utilized adsorbent materials are the Zeolites and activated carbons due to their high efficiency [1].

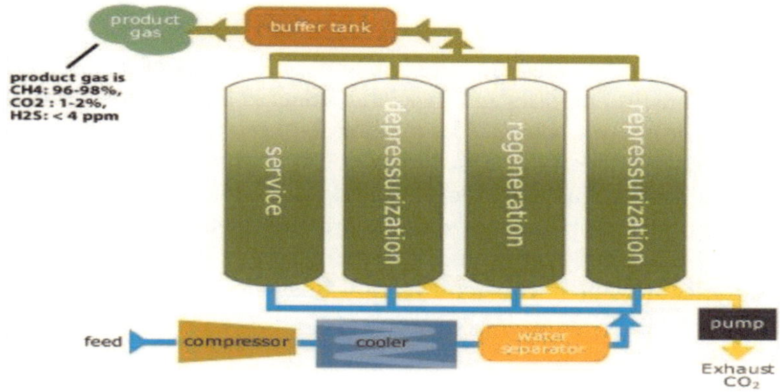

Figure 5.
Pressure swing adsorption system [29].

The pressure swing adsorption process takes place in vertical columns that are packed with absorbents. The process has four steps; adsorption, depressurization, desorption and pressurization in the listed order. Biogas passes through in the pressurized column, while CO_2, N_2, O_2 and H_2S are adsorbed by selected material. Hydrogen sulfide and siloxanes are irreversible adsorbed onto adsorption material and should therefore be removed, together with moisture before injection into the PSA system. It is recommended to use multiple adsorption columns to ensure continuous operation. Once the saturation of adsorbent material is saturated, biogas is allowed into the next column, as regeneration is done for the saturated column. The adsorption column is depressurized to about atmospheric pressure (PSA) or kept under vacuum (VSA). A mixture of CH_4 and CO_2 with high content of CH_4 methane content is released and recycled to the PSA inlet. Biomethane produced can attain purity of 96–98%; but up to 4% CH_4 can be lost within the off-gas stream [10, 25].

The Pressure Swing Adsorption (PSA) technology makes use of the ability of porous adsorbent medium to adsorb specific molecules out of raw biogas and then release through the application of different pressure levels. For the case of raw biogas upgrading [1, 29]. For the biogas upgrading process, the operation is based on the different molecular dimensions of CO_2 which is 0.34 nm, methane CH_4 with 0.38 nm. The application of adsorbent material with cavities of 0.37 nm facilitate retention of CO_2 in the pores, as methane flows out with no retention. The most utilized adsorbent materials are the Zeolites and activated carbons due to their high efficiency [1].

Recent development of the PSA/VSA focus on optimization of adsorption materials and technology. New methods include vacuum swing adsorption system that applies amine-containing nanogel particles supported by carbon fiber having a honeycomb shape whose primary application is the capture of CO_2 from flue gas with potential use in biogas upgrading. In this method, the size of the column and operational costs are reduced by using a rotating design and honeycomb carbon fibers as supportive material, while the combination with amine-containing nanogel particles, increases the recovery of CO_2 [10, 31]. The Amine-containing nanogel particles also reversibly uptake and release CO_2 at lower regeneration temperature of about 75°C which limit the degradation and volatility of amine used [31].

3.4 Absorption techniques

In the absorption techniques, purification and enrichment processes are based on the solubility of constituent gases in biogas in a selected liquid. The commonly used liquids are water or organic solvent like methanol, N-methyl pyrrolidone, and polyethylene and glycol ethers are for absorption of CO in physical absorption plants installations. Amine scrubbing is widely applied for chemical absorption. Water scrubbing has two main applications, namely pre-treatment e.g. before PSA and for the removal of H_2S in actual upgrading. The main limitation of these technique is that it requires significant plant size to achieve high final concentration of methane [30].

3.4.1 Pressurized water scrubbing

Water scrubbing is the most common technology for both biogas cleaning and upgrading. Pressurized water scrubbing depends on the separation of CO_2 and H_2S from raw biogas as a result of increased solubility of CO_2 compared to CH_4. Based on Henry's law, CO_2 solubility in water at 25°C is about 26 times greater than the solubility of methane [13]. Raw biogas is first compressed to 6–10 bar, ND up to 40°C then injected into the absorption column from the bottom side of the tank, while water is supplied from the top while water is supplied from the top side of the column then it flows in the counter-current flow of the gas. The absorption column of the system is filled with random packing material for increased gas-liquid mass transfer [25].

Biomethane is released from the top of the scrubber, the water phase containing the CO_2 and H_2S are circulated to the flush column, in which the pressure is degreased to 2.5–3.5 bar and while traces of CH_4 dissolved in the water is recovered. On the basis of water re-use single pass scrubbing is often employed when water is from sewage treatment plants and "regenerative absorption". Water can be regenerated in a desorption column by decompression at pressure, leading to the removal of CO_2 and H_2S. Water decompression is done by air stripping but where biogas has high concentrations of H_2S, steam or inert are consumed on desorption process to prevent formation of elemental Sulfur by means of air stripping, which leads to operational problems. The regeneration is desirable of huge water requirement by the system e.g. water flow to upgrade 1000 Nm^3 /h of raw biogas needs 180 and 200 m^3/h based on pressure and water temperature. Upon drying, in drying stage, the purity of methane formed can reach 99% purity [13, 25].

Pressurized water scrubbing process works on the basis of the fact that carbon dioxide is more soluble in water than methane. It is the simplest and most popular upgrading technology for biogas. It is necessary to remove H_2S from biogas prior to scrubbing due to its high solubility in water, making its removal difficult. Hence the need to previously remove hydrogen sulfide (H_2S) from biogas, to avoid corrosion and process efficiency reduction. Biogas is first compressed and fed to the absorption column (scrubber), for cooling to (5°C) and pressurized (4–10 bar) to allow water to absorb CO_2 and other impurities. The flash tank is used for water regeneration in the first phase with the recovery of the absorbed biogas, being recycled by injecting at the biogas inlet. The second phase of regeneration takes place in a second column called a stripper through a countercurrent with air, operating under atmospheric pressure [2, 4]. **Figure 6** shows the water scrubbing system.

The main parts of the water scrubbing system as shown in **Figure 6** are the water separator, a compressor a flash tank, desorption column, a cooler, filter, water and an upgraded biogas dryer for upgraded biogas. A water scrubber is a physical scrubber

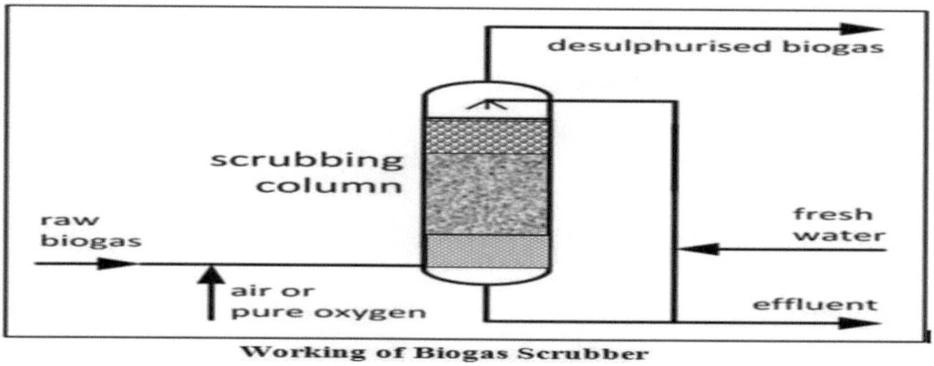

Figure 6.
Water scrubbing system [29].

which exploits the fact that CO_2 is more soluble in water than methane. The CO_2 is separated from the raw biogas and dissolved into the water in the absorption column by application of high pressure of 6–10 bar. The CO2 is then released from the water in the desorption column, by addition of air at atmospheric pressure [29].

3.4.2 Chemical scrubbing

Chemical scrubbing systems use aqueous organic or inorganic compounds to bind the CO_2 or H_2S molecules existing in biogas. The most commonly used scrubbing systems use organic compounds, namely aqueous amine solutions like diethanolamine (DEA), monoethanolamine (MEA) or methyl diethanolamine (MDEA) [25]. Most scrubbing systems using amine solutions have an absorber unit maintained between biogas pressure of 1–2 bars which is injected to the tank bottom with amine solution flowing from the top and a stripper. The solute and solvent (CO_2) undergo a reversible exothermic chemical reaction with the product amine solution which is rich in CO_2 and H_2S which proceeds to the stripper for regeneration operating at a pressure of 1.5–3 bars and temperature of 120-160°C [6].

The heat in the stripper disrupts the chemical bonds formed in absorption phase and which creates steam having CO_2. Upon cooling this steam, the CO_2 is released while the condensate is recirculated back to the stripping column. Some commercial systems can cope with biogas with H_2S content of up to 300 ppm/v, but H_2S poisons the amine, cause corrosion and increase the system energy requirements hence the need to remove H_2S before amine scrubbing [3].

The advantages of the chemical scrubbing using amine solutions include high selectivity of the amines by CO_2 and the substantial extraction compared to other methods e.g. two times more CO_2 per unit volume is absorbed compared with water. The process has significantly low energy requirements mainly because of exothermic reactions and low process pressure operations i.e. 1–2 bar for absorption column and 1.5–3 bar in the stripping column. The draw backs include high energy requirements for solvent regeneration, expensive amine solvents and losses of solvents to evaporation which increase operation costs [3, 25]. **Figure 7** shows that chemical scrubbing using amine solution.

Figure 7 demonstrates a chemical scrubbing system using amine solution. It is equipped with the absorption column a heater cooler and a stripping column with a

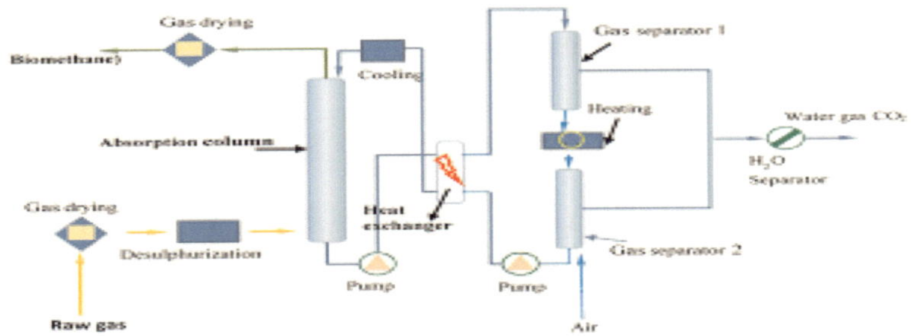

Figure 7.
Chemical scrubbing with amine solution [1].

heating medium being hot water, oil or steam. The removal of CO_2 using reactive systems is not new, but it is less common compared to other technologies like PSA and water scrubbing. The synopsis of features of the chemical scrubbing technology is to use a reagent which chemically binds carbon dioxide molecules for removal from the gas [29].

In chemical scrubbing (CSC), there is reversible reactions between absorbed substances and solvent used. The commonly used biogas upgrading absorption solutions is based on amines i.e. methyl diethanolamine diethanolamine, monoethanolamine, and piperazine. For amine scrubber an absorber tank is used in which carbon dioxide is absorbed from the biogas operating at 20–65°C and 1–2 bar, then followed by a stripper where carbon dioxide is released by heating the stream. Chemical scrubbing with amine facilitates production of high concentration of methane concentration in biomethane greater than CH4 > 99%. The limitation of chemical scrubbing needs pretreatment stage, to remove H_2S and has got high operational and investment costs [30].

The process is similar to pressurized water scrubbing, but is a chemical absorption technique. The solution absorbs CO_2 in biogas, by chemical reaction between amine and CO_2. The absorber is maintained at operating pressure 1–2 bar while the stripper maintained 1.5–3 bar. The process is exothermic, causing temperature rise of amine solution and higher efficiency since the reaction between amine and CO_2 increases with increase in temperature. Hydrogen sulfide (H_2S) should be removed prior to the reaction to avoid poisoning the amine solution [7].

3.4.3 Organic physical scrubbing

Organic physical scrubbing work on the same principle with water scrubbing with the difference being the use of an organic solvent with higher affinity for H_2S and CO_2. Methanol and dimethyl ethers of polyethene glycol (DMPEG) mixtures are all used in biogas upgrading. The process simultaneously absorbs hydrogen sulfide, carbon dioxide, and water due to their higher solubility in polyethene glycol than methane. Examples of commercially available organic physical scrubbing products Selexol® and Genosorb®. These products exhibit high hi solubility of NH_3 and CO_2 compared to H_2O. Selexol® can absorb three times more CO_2 than water hence lower liquid requirements which requires a smaller upgrading [3].

The challenge associated with high solubility of carbon dioxide in organic solvents is difficulty to regenerate organic solvents. Higher solubility of H_2S compared to CO2 in Selexol® leads to increased separation temperatures during the regeneration of the solvent hence higher energy consumption. It is therefore advice able to remove H_2S

before the gas is treated with the solvent. The Selexol process may also be configured to remove H_2S selectively, or non-selectively in order to remove both CO_2 and (H_2S [25]).

In the first stage, raw biogas is compression and cooled to (7–8 bar, 20°C), before injection to the bottom of the absorption column. Since temperature affects Henry's constant the organic solvent is cooled down before it is fed to the column. The desorption column is used to regenerate the organic solvent by heating it to 80°C and reducing pressure to 1 bar. This leads to final methane content of 96–98.5% and less than 2% CH_4 losses, in an optimized full-scale plant [3, 25]. The organic physical scrubbing method is shown in **Figure 8** below.

The main elements of an organic physical scrubbing system as shown in **Figure 8** are the Sulfur absorber, CO_2 absorber, a compressor H_2S concentrator, a reflux pump and accumulator, and stripper reboiler. An organic solvent is used to absorb the CO_2 in raw biogas organic in physical scrubbing method in a process that is theoretically similar to water scrubbing, based on the Henry's law. These solvents include a mix of dimethyl ethers of polyethylene glycol. The relative solubility of the biogas components depends on the solvent used e.g. the solubility of carbon dioxide is much higher in the organic solvent than in water, meaning that the Henry's constant for carbon dioxide is higher. CO_2 has a solubility of 0.18 M/atm in Selexol which is about 3 times higher than in water. CO_2 is about 17 times more soluble than methane in the Genosorb solvent which is a smaller difference than for water, in which CO_2 is 26 times more soluble than methane [29]. These differences in solubility have technical and economic implications.

Concerns over the environment has motivated a shift from the use of conventional solvents to green solvents. This includes the use of deep eutectic solvents (DESs), consisting of two or more components, which are mainly hydrogen bond donors and acceptors [9]. It desirable for the solvents to have a lower melting point, very low vapor pressure and preferably be biodegradable. It is through selection of best fit hydrogen bond donors and hydrogen bond acceptors and donors, that the DESs can be appropriately engineered to yield desired thermodynamic and physical characteristics. It is also possible to remove other biogas contaminants by appropriate process modifications [3, 9].

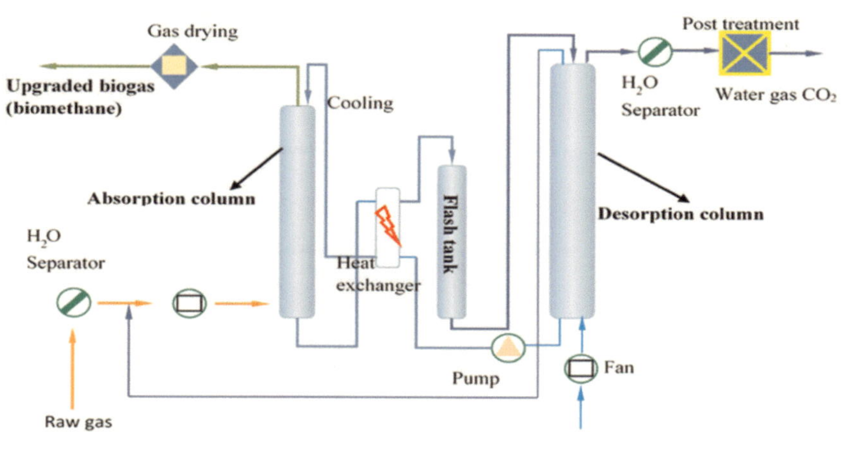

Figure 8.
Organic physical scrubbing [3].

3.5 Membrane separation

One alternative to the conventional absorption in biogas upgrading is membrane technology which make use of membranes which can be made from polymeric materials like cellulose acetate [3]. The membrane is a filter with ability to separate components in raw biogas to the molecular level. Membranes came into use in the 1990s and were initially built with less selective membranes and were applied in applications with lower recovery demand for the methane [29]. The membrane separation technology is a viable alternative to the conventional absorption-based biogas technology. The process relies on the selective permeability properties of membranes which effectively enable separation of biogas components. The various components of raw biogas have different relative permeation rates which can be ordered hierarchically from the slowest to fastest permeation as follows; C_3H_8, CH_4, N_2, H_2S, CO_2 then H_2O [13, 25]. This is demonstrated in **Figure 9** below.

The membrane separation system is demonstrated in **Figure 9** showing stage wise removal of CO_2, O_2, H_2O from raw biogas leaving methane in purified form leaving the membrane. This process can also remove CO_2, H_2O and hydrogen and parts of the oxygen from biogas. The permeation rate depends on the size of molecules hydrophilic through a typical membrane typically made of a glassy polymer [29].

The membranes used in separation are selectively permeable barriers that are designed allow some molecules to pass through but stop or block other with process drivers being the pressure, relative concentration, temperature, and electric charges of the molecules. The membranes used are of three major types, namely polymeric membranes, inorganic membranes, and mixed matrix membranes [2]. Inorganic membranes have got higher mechanical strength, chemical resistance and thermal stability making them more popular. Mixed matrix membranes are the mostly used type of membrane separation in industrial applications. The polymeric and inorganic membrane separation technologies need pre-treatment since H_2S negatively affect medium –term performance. As a strategy to recover up to 99.5% methane, a multi-stage membrane strategy is adopted. The penetration of the membrane technology of separation is high costs and low reliability [2, 30].

The membrane permeation technique works on the basis of the difference in permeability of between the various constituents of biogas. The action of a membrane facilitate separation in which methane is retained, while carbon dioxide and other

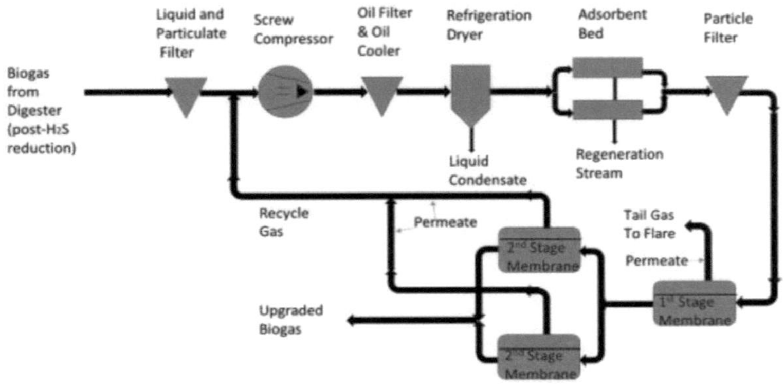

Figure 9.
Operation of membrane separation [29].

constituents penetrate through the membrane. Three types of membranes are currently used for biogas purification; polymeric, inorganic and mixed matrix membranes. The process is not meant to remove H_2S and H_2O, however, they should not be allowed to affect the performance of the membrane. By introducing multiple stages of membranes, CH_4 concentration above 98% and with low operational cost can be attained [3, 4].

From **Figure 10**, it is noted that a gas is fed into a membrane separator where the impurities mainly H_2S, CO_2, are isolated from raw biogas to exceed as permeates while larger methane molecules exceed as retentate i.e. biomethane.

The membrane separation is divided into wet (gas–liquid) and dry (gas–gas) techniques. Biogas is usually pressurized to 20–40 bars or 6–20 resulting in CH_4 abundant gas which will on one side of the membrane with the higher pressure. The CO_2, some H_2S and a significant amount of methane of 10–15% diffuses to the lower pressure side. Contaminants like water, siloxanes, NH_3, VOCs and H_2S are removed before membrane separation to avoid corrosion and clogging. There are different configurations of gas-gas units i.e. single-pass membrane unit or multiple stage membrane units with internal recirculation of permeates and retentates. For one system, about 92% purity of biomethane can be attained while multiple stages can attain 96% or more methane purity [3, 25].

The cold-membrane and cryogenic technologies combination is an interesting approach that can be used in upgrading of biogas. Polyimide and polysulfone membranes can be used in biogas upgrading process to attain up to 98% methane purity, based on simulation results. The process has relatively lower energy requirement of about 1.6 MJ/kg CH_4 which is lower than energy requirements of a standard membrane process which is about 2.4 MJ/kg CH_4. The process can further lower energy requirements to 0.8 MJ/kg CH_4 if the process is coupled with liquefied methane regasification [3, 8].

3.6 Cryogenic separation

Cryogenic separation technology is done by a gradual reduction in the temperature of raw biogas causing liquefaction of CH_4, CO_2 and other constituent parts to ensure methane meets quality standards for Liquefied Natural Gas (LNG). Raw biogas is initially dried and compressed to 80 bars followed by stepwise cooling to −110°C leading to gradual removal of impurities like., siloxanes, H_2O, H_2S, halogens etc. and

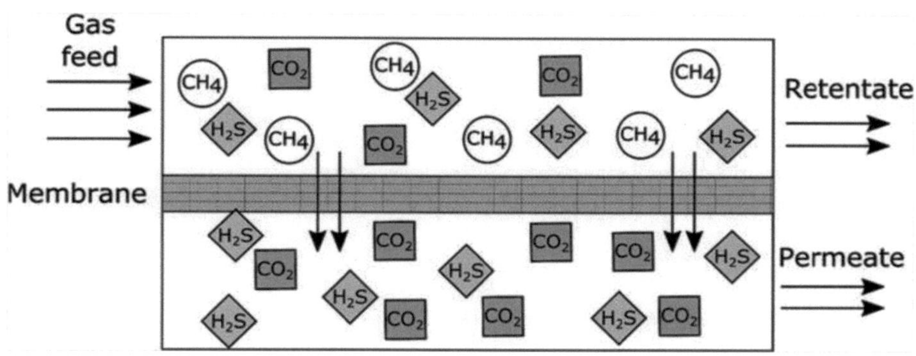

Figure 10.
Principle of membrane separation [6].

CO_2 which is the main impurity in biogas to obtain almost pure biomethane with purity (> 97%) [13, 32].

The physical principle behind cryogenic technique is based on the fact that the gases like carbon dioxide, hydrogen sulfide liquefy and solidify under different pressure and temperature conditions. Therefore, the cryogenic plants operate at very low temperature (−170°C) and high pressure (80 bar). Biogas purification is done by cryogenic technology, with lower methane losses but the process is expensive. The cryogenic process can be used in the production of liquefied natural gas (Bio-LNG) [6, 30].

The cryogenic separation process involves the separation of different gas components based on the basis of their different boiling points by gradually reducing the temperature. The process begins by compressing biogas to 80 bars then reducing the temperature to −25°C. At this temperature and pressure, components that are removed from raw biogas are moisture, halogens, siloxanes and H_2S. Reducing temperature to −55°C, liquefies most of the present CO_2, then further reduction to −85°C is the last step which removes the remaining CO2 in solid form. However, to avoid operational challenges like pipe clogging, ice formation and heat exchanger clogging, the impurities like H_2S, water, siloxanes and halogens are by practice removed prior to cryogenic separation [6].

The purity of biomethane produced with cryogenic upgrading be over 97% with methane losses of less than 2%. The limitation of cryogenic upgrading the high investment and operation costs, methane losses and need for pretreatment to remove impurities. There are additional variants and configurations like cryogenic distillation or cryogenic adsorption [2, 3]. The cryogenic process is shown in **Figure 11**.

From **Figure 11**, we note that the cryogenic separation method is characterized by successive compression and cooling at different pressure to liquefy and isolate the different components of raw biogas. Water is removed in the distillation Colum of the process.

Although the cryogenic separation process is quite promising with interesting performance and results, the method is still under development with just few facilities operating at commercial scale. The process limitations so far are high costs of investment and operation costs, methane losses and clogging derived from increased concentration of solid CO or and presence of other impurities [13, 28].

The cryogenic processes take advantage of the low temperatures to achieve their goals. By allowing component gases in raw biogas to liquify. A process does not have to operate below a fixed temperature level for it to be considered "cryogenic"., but since the However, since the processes involved in this process are done well below

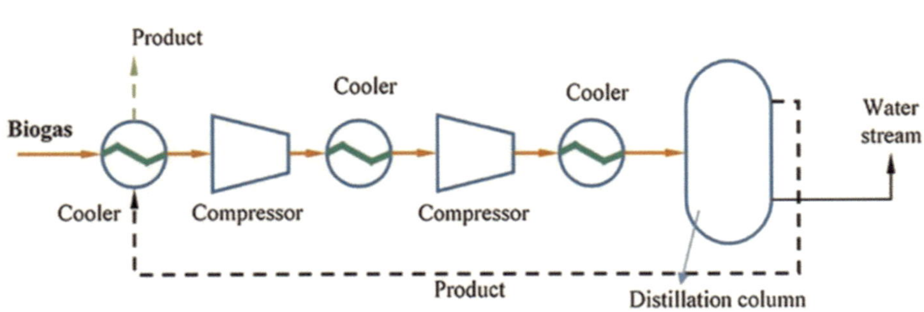

Figure 11.
Cryogenic separation system [6].

−55°C, the common gases in raw biogas can be liquefied and separated which forms the basis for cryogenic biogas upgrading [29].

3.7 Biological upgrade techniques

These processes apply biological separation via hydrogenotrophic methanogenesis consisting of hydrogenotrophic-methanogens to convert CO_2 and H_2 in to CH_4 but commercialization of the processes is limited by several challenges [30]. The positive side is that it requires low investment and operating costs particularly in terms of electricity and head demand. It does not require any chemical products or additional equipment. Easy operation and maintenance. The process leads to high concentrations of hydrogen sulfide while O_2/N_2 excess necessitates additional cleaning while air overload creates an explosive mixture [33].

Although the physicochemical techniques dominate the biogas upgrading market, biological methods have been riding for the last 20 years. Biological technologies for Sulfur treatment in biogas are classified into chemotrophic and photosynthetic types. The advantage of biological techniques is that end products are non-hazardous i.e. sulfur or sulfate with efficiency being same or higher than that of physicochemical technologies while the Sulfur recovered can be used as a raw material for production of sulfuric acid, fungicides and Sulfur fertilizers [33, 34].

3.7.1 Chemotrophic removal of hydrogen sulfide

In this process, the Chemotrophic Sulfur oxidizing bacteria, also known as colorless Sulfur bacteria, is the ideal microbial group used for biodegradation of H_2S. The bacteria are used to oxidize the reduced Sulfur compounds like. Sulfide, polysulfide, elemental Sulfur, thiosulfate, and sulfite to gain chemical energy and utilize CO_2 as a carbon source. Biodegradation of Hydrogen sulfide is done aerobically using where the electron acceptor is O_2 and anaerobically where the electron acceptor is NO_3. The bacteria include genera Thiobacillus, Acidithiobacillus, Sulfolobus, Thiovulum, Thiothrix and Thiospira [3, 24].

3.7.2 Biofiltering

Biofilters remain the simplest type of gas desulphurization systems. The purification system consists of a bioreactor where sulfur-oxidizing bacteria like the Thiobacillus, Pseudomonas and Acidothiobacillus are immobilized on a carrier. In the process, moisturized biogas is injected from the bottom of the biofilter and forced through a moist, packed bed with microbial biofilm which purifies biogas. The bed material is used to supply nutrients or nutrient solution added from the top occasionally. Oxygen whose concentration is 5–10% of volume is supplied by injecting air directly into the gas stream [3, 24]. **Figure 12** shows the biofiltering system.

The biofiltering system as demonstrated in **Figure 12** consists of peristatic pumps for biogas and air, a bioreactor and biogas storage.

Factors influencing the operation of biofilters include the bed medium, moisture content of biogas, gas temperature, the pH, nutrient and oxygen levels, and the development of biofilm. A good or suitable bed material should have large specific area and porosity, create small pressure loss, light in specific weight and cheap. The bed material should absorb gas odor and but retain its nutrients, contain indigenous microorganisms and water i.e. moisture content between 40 and 60%. Suitable bio

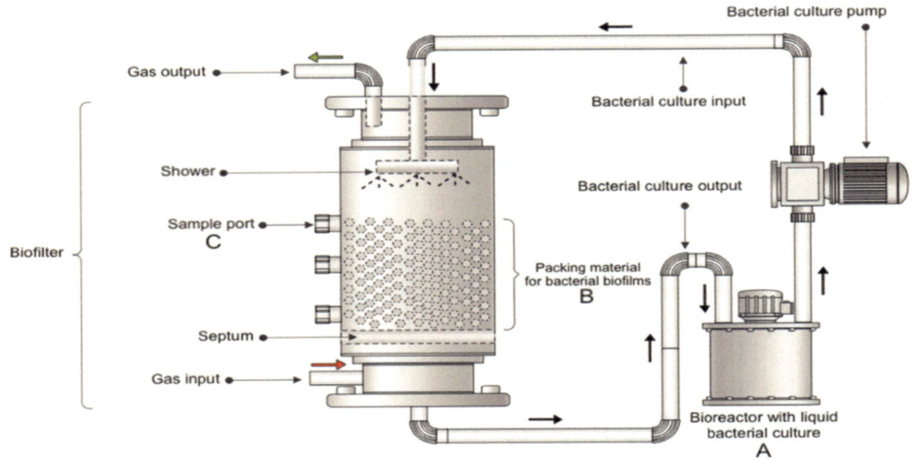

Figure 12.
Biofiltering system [29].

filter materials include natural organic materials like composts, coconut fiber, woodchips/bark, and peats mainly because of their native microorganism consortia and good level of performance [3, 35].

The benefits of using biofilters are reduced operating costs, no chemical requirements. The main limitations in use of bio filters in purification of biogas are media acidification by the sulfuric acid formed from H_2S, degradation and inefficient mixing. Solutions to these challenges include using a carrier with alkaline properties, adding alkaline. Biofilters are also not suitable for high loading rates due to limited buffering capacity and limited control capability for moisture, and pH during high airflows [3, 25].

3.7.3 Bio trickling filters

The general mechanism of biotrickling filtration is same as bio filters, except for the use of inert packing bed material hence the need for continuous supply of the nutrient solution. Plastic supports, activated granular carbon or porous ceramics, are materials commonly used to provide support for biofilm formation. Advantages of bio trickling over traditional filters include better process stability, better control and regulation of the pH and temperature, low flow resistance, less space and continuous nutrient supply. The continuous washout of products of acidic reactions solves the problem of buffering and acidification common in bio filters. However, the challenge of continuous nutrient supply leads to excessive growth of biomass and clogging of anaerobic zones. Commercially available bio trickling systems include BioSulfurex® (DMT Environmental Technology), Biopuric process (Biothane Corporation), Bidox® (Colsen B.V.) and BiogasCleaner® (BioGasclean) [3, 5].

3.7.4 Bio scrubbing

Bioscrubber system is applied in the removal of compounds like ammonia, amines, hydrocarbons, hydrogen sulfide and odorous contaminants. Bio scrubbing system consists of two reactor units with the first reactor as absorption tower where the pollutants are absorbed in a liquid phase before it goes to the second reactor which is activated sludge. Degradation occurs in the activated sludge reactor where microorganisms like Thiobacillus and Thioalkalivibrio) grow in suspended flocks. The effluent generated is recirculated back to the absorption tower. In the removal of H_2S, a sedimentation tank is installed after the second reactor for collection of elemental Sulfur with O_2 being used as the oxidant. Optimal microbial growth and activity are maintained by addition of oxygen, nutrients and pH regulation together with continual purging of by-products and excess biomass out of the system [3, 36].

No injection of N_2 and O_2 is required by bio scrubbers. They are used to handle load fluctuation loads better as well as stable performance due to easy control conditions. The main disadvantage of the bio scrubbing process is high initial costs. Commercially available bio scrubbing systems used for H_2S removal from are THIOPAQ® process and Sulfothane™ which are very similar. In the process, the gas is injected to the absorption tower where counter flow of alkaline solution absorbs H_2S from raw biogas. The formed sulfide-containing goes to a micro-oxygenated reactor. The chemotrophic sulfur oxidizing bacteria, are dominated by haloalkaliphilic Thioalkalivibrio which convert absorbed sulfide to elemental Sulfur [3, 37].

3.7.5 Phototrophic Sulfur removal with anoxygenic bacteria

Phototrophic Sulfur removal uses bacteria with ability to utilize light as an energy source to remove Sulfur compounds from the environment e.g. anoxygenic phototrophic sulfur bacteria, purple non-Sulfur bacteria, cyanobacteria, and phototrophic members of phylum Chloroflexi and Heliobacteria. Some bacteria like the anoxygenic phototrophic sulfur bacteria can oxidize hydrogen sulfide to elemental Sulfur through an oxygenic photosynthesis. Anoxygenic phototrophic Sulfur bacteria consist of two families namely Chlorobiaceae (green sulfur bacteria) and Chromatiaceae (purple Sulfur bacteria). The purple and green sulfur bacteria use light as an energy source, and use reduced Sulfur compounds as electron donors for photosynthetic CO_2 reduction. Sulfide oxidation produces globules of elemental Sulfur. The Chromatiaceae store Sulfur outside of their cells while the Chlorobiaceae store Sulfur inside. The green Sulfur bacteria utilize bacteriochlorophyll *c*, *d*, or *e* found in special light-harvesting organelles (chlorosomes) that allow the growth under the lower intensity light (25–80 lx). The photosynthetic pigments in purple Sulfur bacteria, are bacteriochlorophyll *a* or *b* and various carotenoids i.e. spirilloxanthin, rhodopinal, spheroidene, and okenone) [25].

3.7.6 Chemoautotrophic methods

The chemoautotrophic biogas upgrading methods rely on the hydrogenotrophic methanogens which use H_2 to convert CO_2 to CH_4 based on the following Eq. (1):

$$4H_2 + CO_2 \rightarrow CH4 + H_2O\Delta 0 = -130.7\text{KJ/mol} \tag{1}$$

To make this reaction renewable requires that the source of hydrogen used should be derived from renewable sources hence the need to apply renewable electricity to hydrolyze water for H_2 generation. This facilitate storage of the surplus energy generated by solar and wind to create a new technology called power to gas (P2G). Variable renewable sources need buffering to enable energy delivery when it is dark with no solar and the wind is still. Storage batteries are widely used to store electricity but have drawbacks like capacity limitations, high production cost, and use of toxic materials. Hydrogen (H_2) is a clean energy resource that can be produced by electrolysis of water. As a renewable energy carrier, H_2 has some disadvantages like very low volumetric energy density about 10.88 MJ/m^3 compared to CH_4 which has 36 MJ/m^3 making hydrogen storage a challenge in terms of space requirement. This makes integration of P2G technology for conversion of H_2 to CH_4 really attractive as it integrates wind or solar energy technology as well as biogas technology [25, 37].

Biogas upgrading makes use of existing facilities of the biogas plants which reduces the initial investment cost. The process of chemoautotrophic does not separate or absorb the CO_2, instead it is converted to methane (CH_4) leading to significant increase in the final energy value of the output "wind gas" wind gas is methane produced using the surplus energy from wind turbines) or "solar gas" which is methane produced using surplus solar energy. This technology acts as a precondition for the sustainability of the ambitious biogas development strategy of decoupling the biogas production from the biomass availability. Hydrogen assisted biogas upgrading configurations are classified into in-situ, ex-situ and hybrid designs. The in-situ and ex-situ processes have been experimentally proven with several research undertaken unlike the hybrid concept which is still under development [13, 36].

3.8 In-situ biological biogas upgrading

This method was earlier own presented as one of the raw biogas cleaning technique not aimed at producing biomethane grade biogas. Situ is a biotechnology based on the direct injection of pure air or oxygen and in the process, the bacteria that oxidize H_2S develop with the presence oxygen, leading to the biological removal process of H_2S, to produce sulfur (S) which leaves the digester via the digested. The microorganisms are widely found in the anaerobic environment present in bio-digesters [4, 37]. In situ desorption technology is yet to be fully developed even though it has been around for over 20 years. In-situ is based on the greater solubility of CO_2 over CH_4 in water. The process set up includes an anaerobic digester linked or connected to an external desorption unit. Sludge transported to an aerated desorption column from the digester. Nitrogen or air flowing in counter-current mode dissolves the CO_2 from the sludge in the desorption unit. The sludge desorbed sludge is pumped back into the digester to reabsorb amore CO_2, and the sludge as the sludge is continuously recycled in the desorption column. It is possible to strip out H_2S with dissolved CH_4 and CO_2 from the recirculating sludge by applying large quantities of air or N_2, causing reduction in the H_2S and CO_2 concentration [26, 27].

In the in-situ concept, H_2 is injected into a biogas reactor so that it is coupled with the endogenous CO_2 from anaerobic digestion in the digester for conversion into CH_4 by autochthonous methanogenic archaea. The process can yield methane with purity of 99% if operational parameters like the pH are fully monitored to values above 8.5, as a result of the removal of bicarbonates which inhibits of methanogenesis [13]. CO_2 dissolved in the liquid phase of the reactor dissociates to H^+ and HCO_3^- ions Utilization of carbon dioxide (CO_2) leads to reduction in H^+, which causes concomitant increase in the fermentation pH [13]. The reaction is summarized in Eq. (2) below.

$$H_2O + CO_2 \leftrightarrow H^+ + HCO_3- \qquad (2)$$

Studies on in-situ biogas upgrading reactors show some process inhibition of methanogenesis from bicarbonate consumption which proves the argument that conventional biogas production systems need a pH of 8.5 as the threshold for optimum bio-methanation process for mesophilic and thermophilic activities. Co-digestion with acidic waste can be applied to mitigate the pH increase and alleviate the technical limitations. Another solution to the challenge is application of pH control which enable upgrading to almost pure biomethane [13].

Oxidation of the Volatile Fatty Acids (VFA) and alcohols is thermodynamically feasible in where H_2 concentration is very low. High H_2 levels (> 10 Pa) inhibit bio-digestion and, and promote accumulation of electron sinks like lactate, ethanol, propionate, and butyrate. VFA degradation will cease if there is introduction of sudden and high H_2 concentrations in the reactor which causes system imbalanced or even, fatal deteriorated as a result of excess acidification caused by VFA accumulation. Injection of H_2 in batch reactors at a concentration more than stoichiometric amount for hydrogenotrophic methanogenesis leads to accumulation of acetate, due to stimulated homoacetogenic pathway, and/or decreased methanogenic activity of acetoclastic archaea. However, upon longer term H_2 exposure, there is increase in hydrogenotrophic population which improves the utilization capacity of H_2 and reverts the inhibition [13, 25].

Solubilization of H_2 to the liquid phase is another important parameter since it must cross the interface between the liquid and gas for it to be available for the microorganisms. Hence aqueous solubility of most gasses is rather low, limiting the gas–liquid mass transfer and which retards performance the bioreactor. Therefore, the material and module type used to inject H_2, use of gas recirculation flows and the reactor designs are important aspects of the implementation of sufficient in-situ biogas upgrading. Studies in Batch experiments showed that the rate of uptake of H_2 decreases rapidly at CO_2 concentrations <12% and maximum CH_4 purity attained was 89%. Studies in continuously fed reactors using hollow fiber membranes for H_2 injection in a reactor treating cattle manure and cheese realized 96% CH_4 purity of final gas. Studies in up flow anaerobic sludge blanket reactor, using a hollow fiber membrane in an external degassing unit and realized 94% methane purity [13, 36].

3.9 Ex-situ biological biogas upgrading

Ex- situ biogas upgradation relies on supply of carbon dioxide from external sources and hydrogen in an anaerobic reactor, which eventually contributes to their conversion to methane. The ability of ex situ process to manage high concentrations of influent gases, reduces retention time about 1 hour leading to a smaller device for upgrading. Depending mainly on the reactor used, the ex-situ technology can produce methane with final purity of 79–98%, the main challenge facing this technology is low gas–liquid mass transfer rate [26].

Through studies, it has been established that the operating temperature significantly affects bio-methanation efficiency e.g. enriched thermophilic culture resulted in >60% higher H_2 and CO_2 bioconversion when compared to mesophilic culture in batch. In q typical study, increase of operating temperature from 55 to 65°C showed significant increase in efficient of bio-methanation operation Other than temperature, an adaptation period is needed for microorganisms to efficiently ferment the CO_2 and H_2 gasses e.g. it was established that operating a mesophilic trickle-bed reactor with immobilized hydrogenotrophic culture for 8 months, improved output to CH_4 content of over 96%. Similar results are experienced for bio-methanation efficiency under thermophilic conditions [25].

The reactor type and application of gas recirculation or liquid mixing are important design parameters for biogas upgrading system. Up flow in series or bubble column reactors realize over 98% methane purity, even when H_2 is injected through conventional spargers instead of advanced membrane modules. The trickle bed reactor systems yield higher CO_2 and H_2 conversion efficiency to achieve as high as 98–99% methane purity, due to the formation of biofilm of mixed anaerobic consortia which act as good biocatalyst for the process. High stirring speed or diffusion devices with pore sizes generate gas-bubbles which are able to mix the reactor yield better kinetics and gas quality [13].

The bio filter technology involves passing the biogas through a column having a synthetic material, in the form of a biofilm. The parallel or countercurrent flow maintains the humidity and nutrients, that are essential for the microorganisms that degrade of H_2S [4].

In biological gas scrubber, a two-stage system is used to remove H_2S. In the first stage H_2S scrubbing column, applies sodium hydroxide solution while activated sludge is used in the second stage which is injected with injected with air, because the microorganisms used are aerobic, leading to the solution regeneration [4].

4. Applications of biomethane

Biomethane has superior properties compared to biogas and is attractive substitute of natural gas. It has both industrial and domestic application like use as cooking gas, cogeneration can be packed in containers/cylinders as compressed biomethane and can be injected to natural gas mains for distribution. The main challenge is the cost of processing which is a function of technology used. Bio-CNG, which is a methane-rich compressed fuel in form of biomethane. Bio-CNG is made from pure biogas with more than 97% methane composition pressurized to 20–25 MPa. Compressed bio-CNC is similar in properties to regular CNG in terms of its fuel properties, economy, engine performance, and emissions. Like regular CNG, bio CNG has high octane number, and yields high thermal efficiency. It can therefore substitute the regular compressed natural gas in gas pipelines and other applications including fuel for natural gas power plants [28, 36, 38, 39].

4.1 Hydrogen production

Hydrogen is an ideal raw material for a sustainable energy transformation, but with the challenge being where and how to get hydrogen from renewable sources. Renewable hydrogen can be produced using renewable energy sources and usually produced via water electrolysis [40]. Biogas has applications beyond electricity and biomethane production, as because, through steam reforming, it can be used to manufacture green hydrogen, in a process where a catalyst refines and separates the hydrogen from the gas stream [6, 40]. The most common method used to manufacture hydrogen is by steam-reforming of natural gas, followed by pressure-swing adsorption to remove impurities. However, small reformers like those used for combined heat and power with biogas plants are in commercial operation [40]. The biogas has low heating coefficient due to high composition of carbon dioxide and water vapor interfere with the combustion process [2, 40]. Upon removal of carbon dioxide and water molecules, the methane (CH4) can be used for hydrogen synthesis and bio-fuel production. Methane can be split to hydrogen molecules (H_2) in a process that can be done in a steam/methane reformer. In the process, high pressure and temperature steam is combined with the methane (CH4) to produces flow of hydrogen molecules and CO molecules [2, 6, 33].

It is through thermochemical processes of hydrocarbons that large-scale hydrogen production is manufactured through the reforming process. Biomethane has significant potential application in hydrogen manufacture as a substitute of fossil natural gas as a raw material for reforming processes. The demand for renewable hydrogen production is set to grow significantly due to concerns over fossil fuels depletion and greenhouse gas emissions, and associated concerns over global climate change. The availability of capital, desired hydrogen amount and purity hydrogen and the composition of available biogas will influence the selection of reforming processes [41].

Biomethane as significant application in fuel cell technologies for applications like power generation Fuel cells can use hydrogen to generate electric power just like batteries as well as fuel for the fuel for powering fuel cars. Fuel cell technology in power generation is emission free and hence attractive [42]. Biomethane can be used as a source for renewable hydrogen, for stationary fuel cells and power fuel cell electric vehicles (FCEVs). The Hydrogen-powered FCEVs are environmentally attractive since they have no tailpipe emissions other than making them extremely clean as transport option to fossil fuel powered vehicles [37, 41].

Use of biomethane for hydrogen production can increase energy sustainability can be for energy applications like fossil fuels. Hydrogen can be manufactured by autothermal reforming (ATR), electrolysis or methane reforming (SMR) [43]. Biomethane can be used as a substitute of natural gas which will provide a hedge against growing demand for natural gas [41].

Hydrogen fuel can be used to reduce emissions from engines which are widely used in transportation. Hydrogen fuel cells promise to provide an alternative to internal combustion (IC) engines particularly due to the clean exhaust emissions, renewal nature of the fuel and higher efficiencies. Hydrogen fuel cell vehicles can achieve widespread acceptance except for existing challenges like waste heat removal in mobile applications [44].

4.2 Production of biofuels from biomethane

The transport sector is important since it accounts for about 14% of the global greenhouse gas emissions [45]. Liquefied biomethane is a feasible fuel for power plants and heavy tracks and can also be used as a raw material for production of other fuels and chemicals like methanol, dimethyl ether, and hydrogen fuel. Biomethane is currently used as a transport fuel many countries with benefits of lower environmental impact compared to fossil fuels and several other processed transport fuels [46].

Biofuels include the Bio-CNG which is compressed biomethane similar to (CNG) in properties with industrial, automotive and domestic applications. The process needs removal of impurities likes water, N_2, O_2, H_2S, NH_3 and CO_2 to achieve composition of >97% CH_4, <2% O_2 at 20–25 MPa. Bio-CNG occupies less than 1% of the volume at standard conditions [46, 47].

Biomethane can also be used in the industry as transport fuel by liquefying it to at a high pressure re ranging from 0.5 to 15 MP [4]. In the biological or chemical pathways, biomethane can be converted to methanol, diesel, liquefied petroleum (LPG) and gasoline. Methanol is produced by partial oxidation of methane as shown below;

$$CH_4 + 0.5O_2 \rightarrow CH_3OH \qquad \Delta H^0 = -128\ kJ/mol \tag{3}$$

In another method, methane is biologically converted from to methanol by using methanotrophic bacteria used in methanol production through the action of methane monooxygenase (MMO) enzyme [5].

Methanol can also be produced by reforming methane to syngas then followed by catalytic conversion of syngas to methanol as shown below [6].

$$2H_2 + CO \rightarrow CH_3OH \qquad \Delta H^0 = -91\ kJ/mol \tag{4}$$

$$3H_2 + CO_2 \rightarrow CH_3OH + H_2O \qquad \Delta H^0 = -49\ kJ/mol \tag{5}$$

Methanol can then be converted to gasoline through methanol-to-gasoline process. Biogas or biomethane can be processed to methanol through dry reforming, steam reforming, partial oxidation reforming, autothermal Reforming (ATR) and the Fischer-Tropsch (FT) Process. Synthesis gas (syngas) is the main product of biomethane reforming process. Syngas is a raw material for production of many long chain hydrocarbons [48].

i. Dry reforming

In dry reforming, CO and H_2 are produced by reaction of methane (CH_4) and carbon dioxide (CO_2). The process utilizes two greenhouse gases (CH_4 and CO_2) making it very attractive. Unfortunately, the endothermic reaction reduces the of CO_2 emissions, since CO_2 emitted generate the heat required for the reaction has to be accounted for. Dry reforming is an efficient route for producing synthesis gas yielding a H_2/CO ratio close to 1 [49]. The disadvantage of dry reforming compared with steam reforming it produces lower syngas ratio ($H_2/CO = 1$), The ratio of H_2/CO ratio is influenced by water gas shift reaction (WGS), which reduces the ratio due to reverse reaction that oxidizes hydrogen to water. In this process, the H_2/CO ratio is kept between 1 and 2 by partial oxidation of methane through feeding water. This enhances forward water gas shift reaction. The energy demand by the process is lower since partial oxidation is exothermic [48]. The temperature range for dry reforming process 700–1000°C [48].

$$CH_4 + CO_2 \rightarrow 2CO + 2H_2 \qquad \Delta H^0 = 247\ kJ/mol \tag{6}$$

ii. steam reforming and water shift reaction

This process combines methane in biomethane with water vapor generate CO and H_2 in the in the presence of a catalyst. The process is endothermic and takes place between 650 and 850°C, to produce hydrogen yield of 60–70% [49]. Steam reforming takes place between 700 and 900°C. The two-step chemical reaction is shown below;

$$CH_4 + H_2O \rightarrow CO + 3H_2 \qquad \Delta H^0 = 206\ kJ/mol \tag{7}$$

$$CO + H_2O \rightarrow CO_2 + H_2 \qquad \Delta H^0 = -41\ kJ/mol \tag{8}$$

The process of steam reforming is often followed by a water shift reaction to improve hydrogen generation.

iii. Partial oxidation reforming (POR)

This process is used to produce hydrogen at reduced energy cost because the process is moderately exothermic compared to steam reforming which is highly endothermic. H_2 and CO are produced by the partial oxidation at atmospheric pressure and between 700 and 900°C partial oxidation reforming. The H_2/CO ratio of 2 yield is achieved in full conversion with reduced soot formation. Methane react with oxygen to form carbon dioxide (CO_2) due to decrease in CO selectivity. The combustion is strongly exothermic leading to formation of hot-spots in the reactor bed and coke deposition on the catalyst [49]. In this process, methane is oxidized to syngas as demonstrated below

$$CH_4 + 0.5O_2 \rightarrow CO + 2H_2 \qquad \Delta H^0 = -25.2\ \text{kJ/mol} \tag{9}$$

iv. Autothermal Reforming (ATR)

Autothermal Reforming is a combination of two processes i.e., POR and SR in the presence of carbon dioxide. In Autothermal Reforming (ATR) partial oxidation takes place in the reactor to produce heat needed for steam reforming in the catalytic zone. The process does not need external heating and the reactor is easy to stop and restart. Compared to partial oxidation reaction, the hydrogen yield is higher and consumes less oxygen compared [49].

4.2.1 Upgrading syngas

The Syngas produced from the dry reforming has carbon dioxide which should be removed before it is supplied to the Fischer-Tropsch reactor. Amine absorption has high selectivity for carbon dioxide. Other applications of this technology are separation of CO_2 from flue gases, natural gas cleaning and largescale upgrading of biogas. Common solvents used in the process are alkanolamines like monoethanolamine (MEA), diethanolamine (DEA) or methyldiethanolamine (MDEA) [50, 51].

4.2.2 Fischer-Tropsch (FT) process

The Fischer-Tropsch (FT) synthesis, was name after the German inventors Franz Fischer and Hans Tropsch is a process used to manufacture liquid hydrocarbon fuels like coal-to-liquids (CTL) and/or gas-to-liquids (GTL) based on source of syngas. [50, 51]. The biomethane and natural gas conversion to fuels via Fischer-Tropsch synthesis (FT-synthesis) is feasible at industrial scale [48]. The Fischer-Tropsch (FT) process is coverts syngas to products like LPG, diesel, and jet fuels [24, 50].

The Fischer-Tropsch synthesis (FT-synthesis) polymerizes the carbon and hydrogen atoms in syngas or biomethane to create long chain molecules. The process is run over iron or cobalt catalyst at 20–30 bars [14] in an overall exothermic process leading to polymerization of CH_2 to hydrocarbons with long chains called syncrude. The various reactions in Fischer-Tropsch process are summarized below;

$$(2n + 1)\ H_2 + nCO \rightarrow CnH_{2n+2} + nH_2O \quad (10)$$

$$2nH_2 + nCO \rightarrow C_nH_{2n} + nH_2O \quad (11)$$

Reactors used are include multi-tubular fixed bed, circulating fluidized bed, fixed fluidized bed, and slurry reactor. The reactions for the slurry reactor conditions are 20–30 bar, ands 200–300°C while the syngas H_2/CO ratio of 1–1.8 [48, 50]. For high temperature synthesis, fluidized-bed FT reactors are used to generate light hydrocarbons in form of gaseous hydrocarbons and gasoline and generally have higher output. The catalysts used are Fe and Co which are sensitive to sulfur compounds in syngas [50].

4.2.3 Biofuels from biomethane

Various biofuels can be made from biomethane for the transport sector e.g. methanol, compressed biogas (CBG), hydrogen, liquid biogas (LBG), dimethyl ether and Fischer-Tropsch (FT) fuels [52]. Fuels can be produced by biogas upgrading to biomethane then compressing to make (CBG) or liquefying to make (LBG), or gasification to produce syngas for use in manufacture of hydrogen, methanol, DME and FT diesel [52].

4.3 Biomethane for gas and power grids

In many countries, governments have come up with national support schemes to promote biomethane market. Support mechanisms include feed-in support schemes, green gas products and quota obligations as market drivers in Europe. Biomethane production for many countries is based on organic waste as feedstock, but for Germany which dominates Europe's feed-in market is based on energy crops. In Germany, the main driver is the feed-in tariff for renewable electricity through the Renewable Energy Sources Act (EEG). Biomethane support schemes mainly relay on mass balancing systems or 'book and claim'-certificates. Existing mass balancing systems, can contribute to international market development through creation of common standards. As biomethane becomes popular and relevant in the energy systems, integration into power and gas grids and shift away from subsidies to markets and competition with natural gas will become major issues [53].

Biomethane should meet some standard specifications with respect to storage and transport before it can be practical injection into existing natural gas networks. The presence of various components in different concentrations make it difficult to inject biogas to the grid hence the need for upgrading. Pipeline designers should know the exact he thermodynamic properties of a gas mixture are, particularly in terms of density, and heating value which may tend to vary greatly in biogas [54].

Biomethane has a very important role to play in the transition to renewable sources of energy. Demand based production of biomethane for power generation directly links the gas grid and the electricity grids which can help in balancing the power grid. The gas grid will shift from fossil fuel distribution to provide energy balancing service provider with short-term as well as seasonal storage options. There is increasing integration of decentralized biomethane feed-in into the gas grid for the gas grid infrastructure thus introducing new challenges. There are examples in Germany of facilities that feed in more biomethane to the local distribution network than the total discharge which leads to the need to compress the excess gas and transfer it to a higher level [24, 53].

Production of biomethane from energy crops has a negative impact on agriculture. On the other hand, use of digestion residues as fertilizers close to biogas production site improves local nutrient cycles. Production of energy intensive nitrogen fertilizers and use of declining global phosphorous reserves can be avoided by use of bio-fertilizer from the digesters [8, 18].

Biomethane is the most efficient biofuel in terms of fuel production equivalent per area of crop land needed and is therefore expected to a larger role in the fuel/energy market because of government support, growing use in NGVs and reduction in GHG emissions. There is growing awareness of biomethane and a shift in perception from regarding biomethane as a sub-branch of biomass production to an independent renewable energy resource. And legislation and strategies are recognizing biomethane as an independent energy resource [53].

The main sustainability challenge facing biomethane market is cost of subsidies and need for free market competition with fossil natural gas which can be accelerated if the market price of natural gas rises. The European cap-and-trade for greenhouse gas emissions GHGs is another driving factor for the future. Since use of biomethane omits GHG emissions, there will not be compensation or penalties in form of GHG certificates [37, 53].

The evolution of biomethane markets is expected to create their own demand and supply and also enable and exchange between different countries since the green gas product market open to international trade. Each country has tended to create its own set of biomethane support schemes to address individual situations and are therefore designed with to address the priorities and challenges of specific countries. For biomethane market to grow, countries should open up their support schemes to biomethane imported from neighboring countries to encourage international trade in biomethane [53].

4.4 Electricity from biomethane

Biomethane can be used as fuel for power generation in various prime movers. They include internal combustion engines, gas turbines of varying sizes, fuel cells, among other. The efficiency can be improved through combustion and conversion in set ups like cogeneration and tri-generation schemes [6, 37].

Diesel engines can run on biomethane as a direct substitute of natural gas. Biomethane used can made from biogas upgrading or gasification and methanation schemes [55]. Diesel engines would perform efficiently whether using pure diesel or when running in dual fuel mode as long as the calorific value of fuel is controlled [55, 56].

Electricity from biomethane can be used directly onsite to avoid or limits electricity imports from the grid while excess generated electricity within the design of decentralized power generation systems using a wide range of prime movers for the electric generators e.g. turbines, internal combustion engines, fuel cells, etc. Biomethane can converted to hydrogen fuel for wide renewable applications or used in fuel cells for direct conversion. Various pathways for use of biomethane for power generation are summarized in **Table 4** below.

From **Table 4**, it is biomethane can be used through various conversion technologies with varying characteristics in thermal and electricity generation. The conversion can be done in cogeneration, trigeneration and open conversion systems. Prime movers that can use biomethane include internal combustion engines, gas turbines, fuel cells, and Stirling engines as well as production of fuels for application in transport, heat and electricity generation.

In the transport sector, biomethane has a double role to play in emissions reduction i.e. as a direct fuel substitute of fossil fuels and as feedstock for production of biofuels/

	Device/ technology	Application	Remarks
1	Fuel cell	Generation of electricity	Very efficient, reliable but expensive
2	Hydrogen production	Biomethane can be converted to hydrogen for use as combustion fuel or electricity and process chemicals.	Renewable hydrogen process if its biogas or methane. Hydrogen is manufactured by dry reforming, steam reforming, or hydrolysis
3	Bio-methanation	Biomethane can be fed to natural gas supply as substitute for natural gas	Renewable replacement of fossil natural gas is feasible with use of biomethane
4	Diesel engine	Biomethane can be used as a diesel engine fuel in either dual fuel mode or pure gas engines	Diesel engines have more fuel flexibility and efficiency and can easily use easily use biofuels as fossil fuel substitutes
5	Gas/petrol engine	Biomethane can be used as a fuel for petrol or gasoline engines with little or no modification	Less efficient than diesel engines but are easier to convert to biogas fueled engines.
6	Stirling engine	Stirling engines are also called hot air engines.	Stirling engines have fuel flexibility and can run on a wider range of fuels
7	Gas turbine	Based on size, gas turbines can be micro, small, and large gas turbines in open, closed or combined cycle configuration	Turbines are simple in construction, are versatile and can use raw biogas as well as biomethane and easy to operate.
8	Cogeneration	In cogeneration, biomethane is burn to simultaneously produce useful heat and electricity.	Cogeneration with biomethane as a fuel can be applied on various conversion systems like Stirling engines, diesel engines, gas turbines, hydrogen and fuel cells to increase system efficiency
9	Trigeneration	Trigeneration refers to generation of electricity and both heating and cooling from same fuel/energy resource simultaneously	Trigeneration is the most efficient conversion system but more complex and expensive

Table 4.
Summary of biogas to electricity conversion systems and technologies [2, 6].

chemicals through the Fischer-Tropsch (FT) Process e.g. diesel, jet fuel, and gasoline, and through reforming processes to produce hydrogen and methanol [6, 28].

Biomethane production and use has less environmental impact, but is still associated with some greenhouse gas emissions like CO_2, CH_4 and N_2O whose quantities depend on the technology applied and the source of biogas or feedstock used. Biomethane use can reduce the negative environmental impact and pollution potential while anaerobic digestion and gasification used to produce biogas and syngas can be used for hygiene and of bio wastes further keeping the environment clean and healthy [57–59].

5. Results and discussion

5.1 Summary

Biogas is a product of the anaerobic digestion process with many applications as in generation of renewable energy. The main component of biogas with energy value is methane, but has impurities like moisture, carbon dioxide, siloxanes, hydrogen sulfide, siloxanes, hydrocarbons, oxygen, ammonia, oxygen, carbon monoxide and nitrogen whose presence is undesirable as they reduce the calorific value of biogas and create operational problems in the energy systems. This necessitate biogas cleaning and application of multi-stage technologies to produce upgraded biogas called biomethane. Biomethane gas is a flexible and easy to store fuel with similar properties and applications as natural gas with no need to modify any equipment settings for natural gas devices and equipment.

Technologies that are commercially available for operating biogas upgrading biogas today include amine scrubbers, water scrubbers, PSA units, organic scrubbers and membrane units. Cryogenic upgrading technology though interesting has some important operational challenges that have to be resolved. For medium scale upgrading schemes all the most common upgrading options are feasible. The scrubbing technologies have proved to be effective and efficient and have similar costs of investment and operation. The water scrubber is a preferred choice for many applications due to the simplicity and reliability, but the high purity and very low methane slip from amine scrubbers are notable characteristics. The pressure swing adsorption (PSA) and membrane units, have similar investment costs as the scrubbers. Advances in the membrane technology has also led to low methane slips with this technology, which is a notable progress [1, 6].

The physical and chemical biogas upgrading technologies are the technologies that are currently mature and have reached near optimum technical and economic feasibility. There main limitation is huge energy consumption which limit expansion of the biogas market. Upgrading technology selection may depend on factors like costs, quality of products, location, and technology maturity and requirements [26]. It is possible to attain 98% methane content with a water scrubber but the content of oxygen and nitrogen in the raw biogas cannot be separated in the water scrubber and therefore should be controlled. Oxygen and nitrogen can also be transported with water from the aerated desorption column to the absorption column. With methane composition of 50% in raw biogas, composition of oxygen and nitrogen can double in the final product after upgrading [29, 32].

Biological methods are the less explored methods for industrial-scale testing and optimization scenarios, even show they promise significant potential in terms of techno-economic feasibility with new horizons in hybrid renewable energy applications. There also exist large gaps between pilot scale and commercial scale technology applications like hydrate separation, cryogenic separation, biotechnologies, and chemo lithotrophic-based bioreactors. The main limitation to wide scale use of biomethane is high cost remains a major limitation to widespread biogas application [26].

5.2 Production pathways for biogas

Biomethane is produced by processing raw biogas through multiple steps and by of methane from other components. The various pathways for biogas and biomethane production and applications are summarized in **Figure 13**.

Figure 13, shows various pathways for production of biogas and related by-products, biogas purification, upgrading and various energy applications. The raw materials for biogas production undergo pre-treatment and then fed to the biodigester to undergo anaerobic digestion to produce biogas and digestate. Digestate is applied as green manure, animal bedding, and application like crop irrigation to enhance agricultural production. Raw biogas is purified mainly by desulphurization for

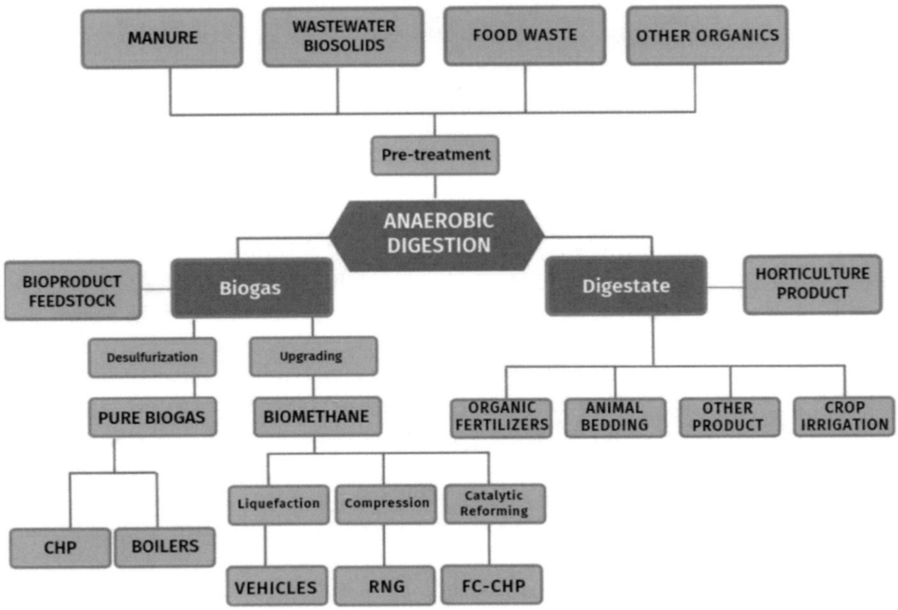

Figure 13.
Pathways for production and application f biogas and biomethane [6].

applications like boilers, cogeneration and engines. Raw biogas can also be upgraded and applied in catalytic reforming for production of hydrogen and biofuels, liquefied to produce bio-liquified natural gas (bio-LNG) or compressed to produce bio-compressed natural gas (bio-CNG) [2, 6, 35].

5.3 Processes comparisons

Biomethane has become an important renewable energy resource for heat and power generation and industrial applications as a feedstock. Applications of biomethane include a transport fuel as a substitute for natural gas, diesel and liquid natural gas, thermal applications i.e. steam and heat generation, combined heat and power, tri-generation, and injection into the natural gas grid upon meeting certain requirements. Biomethane can be manufactured through upgrading of biogas or by gasification followed by methanation process. The approaches in biomethane production have similar efficiencies in biomethane production from the energy output and conservation perspective but since the two technologies have fundamental differences in process and equipment, the cost of the output varies. The main limitations facing biomethane production are the high costs of the process while many biological processes are still under research and development and are yet to be fully commercialized.

Biomethane technology market is a promising venture globally mainly due to existence of mature production, conversion technologies and applications Biomethane remains viable and as a result of abundance in cheap feed stocks supply for anaerobic digestion and gasification. Biomethane has significant flexibility for domestic and industrial scale production and use and is promising to be a leading economical alternative to produce renewable bioenergy.

There are five main biomethane production processes through biogas upgrading. The techniques are pressure swing adsorption which has an option of temperature swing adsorption, absorption technics based on amine, membrane separation, cryogenic separation and biological separation. Biogas upgrading can be significantly improved by combining a wide range of methods ranging from biological and physicochemical processes and adaptation of technologies in the field of advanced oxidation or anaerobic phototrophs. The treatment of biogas can theoretically apply biological methods like chemotrophic or phototrophic to remove H_2S then start upgrading using more efficient physicochemical processes. Purification of biogas is generally a high energy intensive process. But through appropriate choice of a combination of cleaning and upgrading methods based on the methane purity demand saves energy as well as minimize methane loss, in large scale operations. The physicochemical processes are more developed and widely used compared to many biological methods which are still new and not yet commercialized, but they offer significant huge potential in terms of efficiency, feasibility, and technological easiness. Biological methods of upgrading biogas open new horizons for integration of different forms of renewable energy besides electricity, storage advances and decoupling bioenergy production from the availability of biomass resources. The various processes have different benefits and limitation. The water scrubbing can simultaneously remove CO_2, H_2S, NH_3 and dust but consumes a lot of water. On the other hand, the pressure swing adsorption technique requires biogas pretreatment since water and hydrogen sulfide can damage the adsorbents but, has low energy requirement. The advantage of Amine absorption is low methane loss and produces high quality CO_2, although it has high energy consumption. The advantage of the membrane permeation systems is there compactness and ease of operation but yields a relatively low methane (CH_4) purity.

Biomethane can be used in power generation using prime movers gas turbines, micro turbines, diesel engines, petrol engines, Stirling engines. Other applications are hydrogen production, manufacture of transport fuels, fuel for cogeneration and trigeneration, compression to bio-CNG and LPG, syngas production, methanol production.

5.4 Applications

Biogas has multiple applications which include heat and electricity generation. Electricity can be produced from biogas at sewage works, by means of a combined heat and power (CHP) engine, gas turbines, petrol engines, modified diesel engines of dual engines, among others. Biogas and biomethane can also be used as a fuel in automobiles to power an internal combustion engine or a fuel cell in cleaner processes compared to use of fossil fuels [2, 5, 6, 60].

Upgraded biogas or biomethane can attain same properties as natural gas and hence be used as a substitute fuel for natural gas as green natural gas. Biomethane can be injected to natural gas pipelines for use in applications domestic heating and cooking, power generation and feedstock for many industrial processes [6, 35, 61].

6. Conclusions

Biogas is a product of the anaerobic digestion process with many applications as in generation of renewable energy. The main component of biogas with energy value is methane, but has impurities like moisture, carbon dioxide, siloxanes, hydrogen sulfide, siloxanes, hydrocarbons, oxygen, ammonia, oxygen, carbon monoxide, and nitrogen whose presence is undesirable as they reduce the calorific value of biogas and create operational problems in the energy systems. This necessitates biogas cleaning and upgrading to biomethane. Biomethane gas is a flexible and easy to store as a fuel having similar properties and applications as natural gas with little or no modifications to the natural gas equipment.

Biogas upgrading methods can be classified into physical, chemical and biological methods like water scrubbing, physical absorption, pressure swing adsorption, cryogenic separation, membrane separation, chemical scrubbing, chemoautotrophic methods, photosynthetic upgrading and desorption. The physical and chemical upgrading technologies have almost reached optimal level but still have high energy requirements. High-pressure water scrubbing is more economic for small-sized plants, but potassium carbonate scrubbing has high net value for large-sized plants. Therefore, physicochemical methods are technologically ready compared to biological methods which are still new and not yet commercially available, although they offer huge potential in respect to feasibility, technological easiness, and potential. Through biological upgrading new opportunities for integrating different forms of renewable energy are availed besides upgrading including electricity storage advances and decoupling bioenergy production from availability of biomass.

Biogas can be cleaned or purified to remove harmful components like moisture and H_2S without necessarily upgrading to biomethane which is mainly about the removal of Biogas. removal of CO_2. Some upgrading methods remove other impurities in addition to CO_2, while others require upfront removal of H_2O and H_2S. Raw biogas cleaning/treatment and upgrading which enables the use of biogas in applications like vehicles fuel or for injection into the natural gas grid as a substitute for natural gas. There have been significant developments over the last few years, in the field of biogas cleaning and upgrading through process improvements and development of new technologies although water scrubbing, PSA and amine scrubbing currently dominate the market. Membrane separation is a technology is while organic physical scrubbers have limited share of biogas upgrading market. Cryogenic upgrading technologies, which are potentially the best choice for combination with liquefaction of biomethane, still face operational challenges that may be resolved.

The market for biomethane globally is promising mainly due to existence of mature production, energy conversion technologies and applications and abundance of cheap feed stocks for anaerobic digestion and gasification. Biomethane has significant flexibility for domestic and industrial scale production and use and is promising to be a leading economical alternative to natural gas.

The main limitations facing biomethane production are high costs of the process while others like biological techniques are still under development. Biogas upgrading can be significantly improved by combining a wide range of methods ranging from biological and physicochemical processes and adaptation of technologies in the field of advanced oxidation or anaerobic phototrophs. The treatment of biogas can theoretically apply biological methods like chemotrophic or phototrophic to remove H_2S then start upgrading using more efficient physicochemical processes. Purification of biogas is generally a high energy intensive process. But through appropriate choice of a combination of cleaning and upgrading methods based on the methane purity demand saves energy as well as minimize methane loss, in large scale operations. The physicochemical processes are more developed and widely used compared to many biological methods which are still new and not yet commercialized, but they offer significant huge potential in terms of efficiency, feasibility, and technological easiness. Biological methods of upgrading biogas open new horizons for integration of different forms of renewable energy besides electricity, storage advances and decoupling bioenergy production from the availability of biomass resources.

Biomethane can be used in power generation using various available uses like the use of gas turbines, micro-turbines, diesel engines, petrol engines, Stirling engines besides thermal applications as a biofuel for transport and industrial applications. The tracks for production of fuels biomethane are compressing to produce (CBG) or liquefying to make (LBG), hydrogen production, methanol, production, DME, and FT diesel.

Biomethane can be manufactured through upgrading of biogas or by gasification followed by methanation process. The approaches in biomethane production have similar efficiencies in biomethane production from the energy output and conservation perspective but since the two technologies have fundamental differences in process and equipment, the cost of the output varies. High initial and operating costs remain the limiting factor facing the biomethane technology market, while several promising technologies are still under research and development.

Acknowledgements

The authors wish to appreciate researchers and scholars in the field of geothermal energy and electricity for providing significant, credible, and reliable information in all aspects of geothermal energy with ready access. This made the production of this study successful.

Funding

There is no funding provided for this research and the whole exercise was fully funded the researchers.

Author's contribution

The first author conceptualized the manuscript and produced the draft for review by the second author who also facilitated funding for publication of the manuscript.

Availability of data

The research has provided all data and information used and did not use any undeclared data and information. However, any datasets used and/or analyzed during the current study are available from the corresponding author on reasonable request.

Conflict interest

The authors declare that they have no conflict of interest.

Consent for publication

The authors have authority to publish the research work.

Ethical approval and consent to participate

Not applicable.

Author details

Moses Jeremiah Barasa Kabeyi* and Oludolapo Akanni Olanrewaju
Industrial Engineering Department, Durban University of Technology,
Durban, South Africa

*Address all correspondence to: 2206469@dut4life.ac.ke; moseskabeyi@yahoo.com

References

[1] Ardolino F, Cardamone GF, Parrillo F, Arena U. Biogas-to-biomethane upgrading: A comparative review and assessment in a life cycle perspective. Renewable and Sustainable Energy Reviews. 2022;**139**:110588. DOI: 10.1016/j.rser.2020.110588

[2] Kabeyi MJB, Olanrewaju OA. Technologies for biogas to electricity conversion. Energy Reports. 2022;**8** (Supplement 16):774-786. DOI: 10.1016/j.egyr.2022.11.007

[3] Struk M, Kushkevych I, Vítězová M. Biogas upgrading methods: Recent advancements and emerging technologies. Reviews in Environmental Science and Bio/Technology. 2020;**19**(3): 651-671. DOI: 10.1007/s11157-020-09539-9

[4] Domingues PS, Pala H, Oliveira NS. Main biogas upgrading technologies. International Journal of Environmental Science and Natural Resources. 2021; **27**(4):556219. DOI: 10.19080/IJESNR.2021.27.556219

[5] Kabeyi M, Olanrewaju O. Diesel to gas engine power plant conversion: A review and preliminary design for an operting power plant, (in en). Journal of Energy Management and Technology. 2023;**7**(2): 103-115. DOI: 10.22109/jemt.2022.292982.1312

[6] Kabeyi MJB, Olanrewaju OA. Biogas production and applications in the sustainable energy transition. Journal of Energy. 2022;**2022**(8750221):43. DOI: 10.1155/2022/8750221

[7] Kabeyi MJB, Olanrewaju OA. Energy and environment: Opportunities and challenges for the energy transition. In: Presented at the Fifth European Conference on Industrial Engineering and Operations Management, Rome, Italy, July 26–28, 2022. Michigan, US: IEOM Society International; 2022. p. 427. [Online] Available from: https://ieomsociety.org/proceedings/2022rome/427.pdf

[8] Kabeyi MJB, Olanrewaju O. Optimum biodigestor design and operations. In: Presented at the Fifth European Conference on Industrial Engineering and Operations Management, Rome, Italy, July 26–28, 2022. Michigan, US: IEOM Society International; 2022. p. 424. [Online] Available from: https://ieomsociety.org/proceedings/2022rome/424.pdf

[9] Abatzoglou N, Boivin S. A review of biogas purification processes. Biofuels, Bioproducts and Biorefining. 2009;**3**(1): 42-71. DOI: 10.1002/bbb.117

[10] Kadam R, Panwar NL. Recent advancement in biogas enrichment and its applications. Renewable and Sustainable Energy Reviews. 2017;**73**: 892-903. DOI: 10.1016/j.rser.2017.01.167

[11] Kabeyi MJB, Oludolapo AO. Viability of wellhead power plants as substitutes of permanent power plants. In: Presented at the 2nd African International Conference on Industrial Engineering and Operations Management, Harare, Zimbabwe, December 7–10, 2020. Michigan, US: IEOM Society International; 2020. p. 77 [Online]. Available from: http://www.ieomsociety.org/harare2020/papers/77.pdf

[12] Kabeyi MJB, Olanrewaju OA. The potential of grid power generation from municipal solid waste for Nairobi city. In: Presented at the 2nd African International Conference on Industrial Engineering and Operations Management, Harare, Zimbabwe, 5–7

December 2020. Michigan, US: IEOM Society International; 2020. p. 081 [Online]. Available from: http://ieomsociety.org/harare2020/papers/81.pdf

[13] Angelidaki I et al. Biogas upgrading and utilization: Current status and perspectives. Biotechnology Advances. 2018;**36**(2):452-466. DOI: 10.1016/j.biotechadv.2018.01.011

[14] Kabeyi M, Olanrewaju O. Optimum Bio-digester Design and Operations. In: Presented at the Fifth European International Conference on Industrial Engineering and Operations Management, Rome, Italy, July 26–28, 2022. Michigan, US: IEOM Society International; 2022 [Online]. Available from: https://ieomsociety.org/proceedings/2022rome/424.pdf

[15] Kabeyi M, Olanrewaju O. Diesel powerplants: Design and operation and performance enhancements. In: Presented at the Fifth European Conference on Industrial Engineering and Operations Management, Rome, Italy, July 26–28, 2022. Michigan, US: IEOM Society International; 2022. p. 425. Available from: https://ieomsociety.org/proceedings/2022rome/425.pdf

[16] Kabeyi M, Olanrewaju O. Performance analysis and development of an export cogeneration plant for a 3000 TCD sugar cane factory. In: Presented at the 12th Annual Istanbul International Conference on Industrial Engineering and Operations Management, Istanbul, Turkey, March 7–10, 2022. Michigan, USA: IEOM Society International; 2022. p. 409. Available from: https://ieomsociety.org/proceedings/2022istanbul/409.pdf

[17] Kabeyi MJB, Oludolapo AO. Optimization of biogas production for optimal abattoir waste treatment with bio-Methanation as solution to Nairobi slaughterhouses waste disposal. In: Presented at the 2nd African International Conference on Industrial Engineering and Operations Management, Harare, Zimbabwe, 5–7 December 2020. Michigan, US: IEOM Society International; 2020. p. 083 [Online]. Available from: http://ieomsociety.org/harare2020/papers/83.pdf

[18] Kabeyi MJB, Oludolapo AO. Development of a biogas plant with electricity generation, heating and fertilizer recovery systems. In: Presented at the 2nd African International Conference on Industrial Engineering and Operations Management, Harare, Zimbabwe, 5–7 December 2020. Michigan, US: IEOM Society International; 2020. p. 082 [Online]. Available from: http://ieomsociety.org/harare2020/papers/82.pdf

[19] Kabeyi MJB, Oludolapo AO. Performance analysis of an open cycle gas turbine power Plant in Grid Electricity Generation. In: Presented at the 2020 IEEE International Conference on Industrial Engineering and Engineering Management (IEEM), Singapore, Singapore, 14–17 December 2020, IEEM20-P-0438. Michigan, US: IEOM Society International; 2020 [Online]. Available from: https://ieeexplore.ieee.org/stamp/stamp.jsp?tp=&arnumber=9309840

[20] Kabeyi M, Olanrewaju O. Performance analysis of a sugarcane bagasse cogeneration power plant in grid electricity generation. In: Presented at the 11th Annual International Conference on Industrial Engineering and Operations Management Singapore, March 7–11, 2021. Michigan, US: IEOM Society International; 2021 [Online]. Available from: http://www.ieomsociety.org/singapore2021/papers/201.pdf

[21] Kabeyi M, Olanrewaju O. Development of a cereal grain drying system using internal combustion engine waste heat. In: Presented at the 11th Annual International Conference on Industrial Engineering and Operations Management Singapore, March 7–11, 2021. Michigan, US: IEOM Society International; 2021 [Online]. Available from: http://www.ieomsociety.org/singapore2021/papers/188.pdf

[22] Kabeyi MJB, Olanrewaju OA. The role of electrification of transport in the energy transition. In: Presented at the Fifth European Conference on Industrial Engineering and Operations Management, Rome, Italy, July 26–28, 2022. Michigan, US: IEOM Society International; 2022. p. 426 [Online]. Available from: https://ieomsociety.org/proceedings/2022rome/426.pdf

[23] Kabeyi MJB, Olanrewaju AO. Application of geothermal wellhead generators in sustainable power generation. Geothermal Resources Council Transactions. 2022;**46**: 1692-1718. Art no. 1034703 Available from: https://www.geothermal-library.org/index.php?mode=pubs&action=view&record=1034703

[24] Yu T, Chen Z, Liu Z, Xu J, Wang Y. Review of hydrogen sulfide removal from various industrial gases by zeolites. Separations. 2022;**9**(9):1-17. DOI: 10.3390/separations9090229

[25] Pudi A, Rezaei M, Signorini V, Andersson MP, Baschetti MG, Mansouri SS. Hydrogen sulfide capture and removal technologies: A comprehensive review of recent developments and emerging trends. Separation and Purification Technology. 2022;**298**:121448. DOI: 10.1016/j.seppur.2022.121448

[26] Ahmed SF et al. Biogas upgrading, economy and utilization: A review. Environmental Chemistry Letters. 2021; **19**(6):4137-4164. DOI: 10.1007/s10311-021-01292-x

[27] Abou Rjeily M, Gennequin C, Pron H, Abi-Aad E, Randrianalisoa JH. Pyrolysis-catalytic upgrading of bio-oil and pyrolysis-catalytic steam reforming of biogas: A review. Environmental Chemistry Letters. 2021;**19**(4): 2825-2872. DOI: 10.1007/s10311-021-01190-2

[28] Kabeyi MJB, Olanrewaju OA. Conversion of diesel and petrol engines to biogas engines as an energy transition strategy. In: Presented at the 4th African International Conference on Industrial Engineering and Operations Management, Nsukka, Nigeria, April 5–7, 2022. Michigan, US: IEOM Society International; 2022. p. 448 [Online]. Available from: https://ieomsociety.org/proceedings/2022nigeria/448.pdf

[29] Bauer F, Hulteberg C, Persson T, Tamm D. Biogas upgrading – Review of commercial technologies, Svenskt Gastekniskt Center, Sweden. SGC Rapport. 2013;**270** Accessed: 20 May 2023. [Online]. Available from: http://vav.griffel.net/filer/C_SGC2013-270.pdf

[30] Prussi M, Padella M, Conton M, Postma ED, Lonza L. Review of technologies for biomethane production and assessment of Eu transport share in 2030. Journal of Cleaner Production. 2019;**222**:565-572. DOI: 10.1016/j.jclepro.2019.02.271

[31] Gao J, Hoshino Y, Inoue G. Honeycomb-carbon-fiber-supported amine-containing nanogel particles for CO2 capture using a rotating column TVSA. Chemical Engineering Journal. 2020;**383**:123123. DOI: 10.1016/j.cej.2019.123123

[32] Kabeyi MJB, Olanrewaju OA. Bagasse electricity potential of conventional sugarcane factories. Journal of Energy. 19 May 2023;**2023**:1-25. DOI: 10.1155/2023/5749122

[33] Karuppiah T, Azariah VE. Biomass Pretreatment for enhancement of biogas production. In: Anaerobic Digestion. London: IntechOpen; 2019

[34] Kabeyi MJB, Olanrewaju OA. A techno-economic assessment of diesel to gas power plant conversion. In: Presented at the 12th Annual Istanbul International Conference on Industrial Engineering and Operations Management, Istanbul, Turkey, March 7–10, 2022. Michigan, US: IEOM Society International; 2022. p. 406 [Online]. Available from: https://ieomsociety.org/proceedings/2022istanbul/406.pdf

[35] Kabeyi M, Olanrewaju O. Slaughterhouse waste to energy in the energy transition with performance analysis and design of slaughterhouse biodigestor, (in en). Journal of Energy Management and Technology. 2022; **6**(3):188-208. DOI: 10.22109/jemt.2021.292954.1309

[36] Kabeyi M, Olanrewaju O. Feasibility of conversion from diesel engine to natural gas power plants. In: Presented at the IECON 2022 – 48th Annual Conference of the IEEE Industrial Electronics Society, Brussels, Belgium, 17–20 Oct. 2022. New York City, US. 2022 [Online]. Available from: https://ieeexplore.ieee.org/document/9968428

[37] Ou H-W, Chou M-S, Chang H-Y. Removal of hydrogen Sulfide from biogas using a bubbling tank fed with aerated wastewater. Aerosol and Air Quality Research. 2020. DOI: 10.4209/aaqr.2019.12.0647

[38] Kour D et al. Technologies for Biofuel Production: Current Development, Challenges, and Future Prospects. Cham: Springer; 2019

[39] Kabeyi MJB, Olanrewaju OA. Conversion from diesel to dual fuel power generation and implications on the transition. In: Presented at the 7th North American International Conference on Industrial Engineering and Operations Management Orlando, Florida, USA June 12–14, 2022, Conference Paper. Michigan, US: IEOM Society International; 2022. p. 356 [Online]. Available from: https://ieomsociety.org/proceedings/2022orlando/356.pdf

[40] A. Admin. Green hydrogen from biogas. H2 International. 2022. Available from: https://www.h2-international.com/2020/09/15/green-hydrogen-from-biogas/ [accessed 1 July 2022]

[41] Saur G, Milbrandt A. Renewable Hydrogen Potential from Biogas in the United States. Denver West Parkway, USA: NREL of the U.S. Department of Energy; 2014 [Online]. Available from: https://www.nrel.gov/docs/fy14osti/60283.pdf

[42] NaturalGas.org. Natural gas and the environment. NaturalGas.org. Available from: http://naturalgas.org/environment/naturalgas/ [accessed 2021]

[43] Antonini C, Treyer K, Streb A, Mijndert VDS, Bauer C, Mazzo M. Hydrogen production from natural gas and biomethane with carbon capture and storage – A techno-environmental analysis. Sustainable, Energy & Fuels. 2020;**2020**(4):2967. DOI: 10.1039/D0SE00222D

[44] Fly A, Thring RH. System thermal and water balance in an evaporatively cooled PEM fuel cell vehicle. In: Vehicle

Thermal Management Systems Conference Proceedings (VTMS11). Woodhead Publishing; 2013. pp. 267-277

[45] Boden TA, Marland G, Andres RJ. Global, Regional, and National Fossil-Fuel CO2 Emissions, 26 September 2017. Oak Ridge, Tenn., U.S.A: U.S. Department of Energy; 2017 [Online]. Available from: https://cdiac.ess-dive.lbl.gov/trends/emis/overview_2014.html

[46] Lyng K-A, Brekke A. Environmental life cycle assessment of biogas as a fuel for transport compared with alternative fuels. Energies. 2019;**12**(3):532 [Online]. Available from: https://www.mdpi.com/1996-1073/12/3/532

[47] Li Y. Converting Biogas to Transportation Fuels, Apr 24, 2014. Ohio, USA: Ohio State University; 2014 [Online]. Available from: https://ohioline.osu.edu/factsheet/AEX-653.2

[48] Ashraf MT, Schmidt JE, Bastidas Oyanedel JR. Conversion efficiency of biogas to liquids fuels through Fischer-Tropsch process. In: Presented at the 23rd European Biomass Conference and Exhibition, 2015. Odense, Denmark. 2015 [Online]. Available from: http://www.etaflorence.it/proceedings/?detail=11205

[49] Alves HJ, Bley Junior C, Niklevicz RR, Frigo EP, Frigo MS, Coimbra-Araújo CH. Overview of hydrogen production technologies from biogas and the applications in fuel cells. International Journal of Hydrogen Energy. 2013;**38**(13):5215-5225. DOI: 10.1016/j.ijhydene.2013.02.057

[50] National Energy Technology Laboratory. 10.2. Fischer-Tropsch synthesis. U.S. Department of Energy National Energy Technology Laboratory. Available from: https://www.netl.doe.gov/research/coal/energy-systems/gasification/gasifipedia/ftsynthesis; [accessed 20221]

[51] Herz G, Reichelt E, Jahn M. Design and evaluation of a Fischer-Tropsch process for the production of waxes from biogas. Energy. 2017;**132**:370-381. DOI: 10.1016/j.energy.2017.05.102

[52] Dahlgren S. Biogas-based fuels as renewable energy in the transport sector: an overview of the potential of using CBG, LBG and other vehicle fuels produced from biogas. Biofuels. 28 May 2022;**13**(5):587-599. DOI: 10.1080/17597269.2020.1821571

[53] Stephan B. 19 - market development and certification schemes for biomethane. In: Wellinger A, Murphy J, Baxter D, editors. The Biogas Handbook. Woodhead Publishing; 2013. pp. 444-462

[54] Farzaneh-Gord M, Mohseni-Gharyehsafa B, Arabkoohsar A, Ahmadi MH, Sheremet MA. Precise prediction of biogas thermodynamic properties by using ANN algorithm. Renewable Energy. 2020;**147**:179-191. DOI: 10.1016/j.renene.2019.08.112

[55] Mitzlaff K. Engines for Biogas. Eschborn, Germany. 3-528-02032-6: Vieweg & Sohn Verlagsgesellschaft mbH; 18 December 2020. 1988 [Online]. Available from: https://biogas.ifas.ufl.edu/ad_development/documents/Engines%20for%20biogas.pdf

[56] Mungut M, Bore J, Kakinyi M. Housing characteristics and household population: Findings from the Kenya demographic and health survey 2014. Pan African Medical Journal. 2018;**30**. DOI: 10.11604/pamj.2018.30.166.1477

[57] Mathur A, Verma L. Management and Utilization of Biogas Plant Slurry.

Udaipur: Himanshu Publications; 1988 [Online]. Available from: https://www.econbiz.de/Record/management-and-utilization-of-biogas-plant-slurry-mathur/10000778141

[58] Li T, Tang D, Li Z, Du J, Zhou T, Jia Y. Development and test of a Stirling engine driven by waste gases for the micro-CHP system. Applied Thermal Engineering. 2012;**2012**:33-34. DOI: 10.1016/j.applthermaleng.2011.09.020

[59] Bhardwaj S, Das P, Review A. Advantages and disadvantages of biogas. International Research Journal of Engineering and Technology (IRJET). 2017;**4**(10):890-893

[60] Kabeyi MJB, Olanrewaju OA. Cogeneration potential of an operating diesel engine power plant. Energy Reports. 2022;**8**(16):744-754. DOI: 10.1016/j.egyr.2022.10.447

[61] Kabeyi MJB, Olanrewaju OA. Sugarcane molasses to energy conversion for sustainable production and energy transition. In: Presented at the 12th Annual Istanbul International Conference on Industrial Engineering and Operations Management, Istanbul, Turkey, March 7–10, 2022. Michigan, US: IEOM Society International; 2022. p. 405 [Online]. Available from: https://ieomsociety.org/proceedings/2022istanbul/405.pdf

Chapter 2

Structure and Sludge-Water Mixing of Anaerobic Reactor

Yang Min

Abstract

Complete sludge-water mixing reaction is a common mode of operation for anaerobic reactors. However, sludge-water mixing must be carried out at low mixing intensities or flow velocity in order to maintain stable anaerobic colonies in the anaerobic sludge and thus achieve high anaerobic biochemical reaction efficiencies. In this chapter, the Continuous Stirring Tank Biochemical Reactor (CSTBR) was defined in terms of the sludge-water mixing time scale $\bar{t}$ and the biochemical reaction time scale T ($\bar{t} << T$) for the analysis of sludge-water mixing reaction of anaerobic reactor, it tends to CSTR when $\bar{t} \rightarrow 0$. Upflow Anaerobic Sludge Blanket (UASB) reactors and Internal Reflux Packed-Bed Anaerobic Reactors (IRPAR) with confined anaerobic sludge can mix sludge and water with lower mixing intensities to achieve the CSTBR mode, the corresponding velocity can be as low as 0.4–5.0 m/h. This chapter analyses the structural and operational characteristics of these two types of anaerobic reactors, and presents the corresponding sludge-water mixing calculation models and the operation conditions required to achieve the CSTBR mode. Such mixing model analysis is an effective way of designing structural of anaerobic reactors and controlling operation.

Keywords: anaerobic reactor, CSTBR, sludge, sludge-water mixing, mixing time scale

1. Introduction

1.1 Anaerobic reactor

The anaerobic reactor is a biochemical reactor that efficiently carries out anaerobic biochemical degradation of organic matter. The biochemical reaction of organic matter degradation by anaerobic colonies in an anaerobic reactor is carried out by the activity of microbial symbiont, which includes a complex process consisting of many species of bacteria and a number of intermediate steps. If the organic matter is made up of complex organic compounds. It first needs to be hydrolyzed into simple organic matter, which is then fermented by acid-producing bacteria to produce volatile acids, which are then converted into acetate and hydrogen by specialized hydrogen-producing acetic acid-producing bacteria. Finally, methanogenic bacteria convert the acetate and hydrogen into methane. This related, complex process is shown in **Figure 1**.

As can be seen from the tandem reaction route in **Figure 1**, the anaerobic degradation process can proceed efficiently when the microorganisms in each sequence

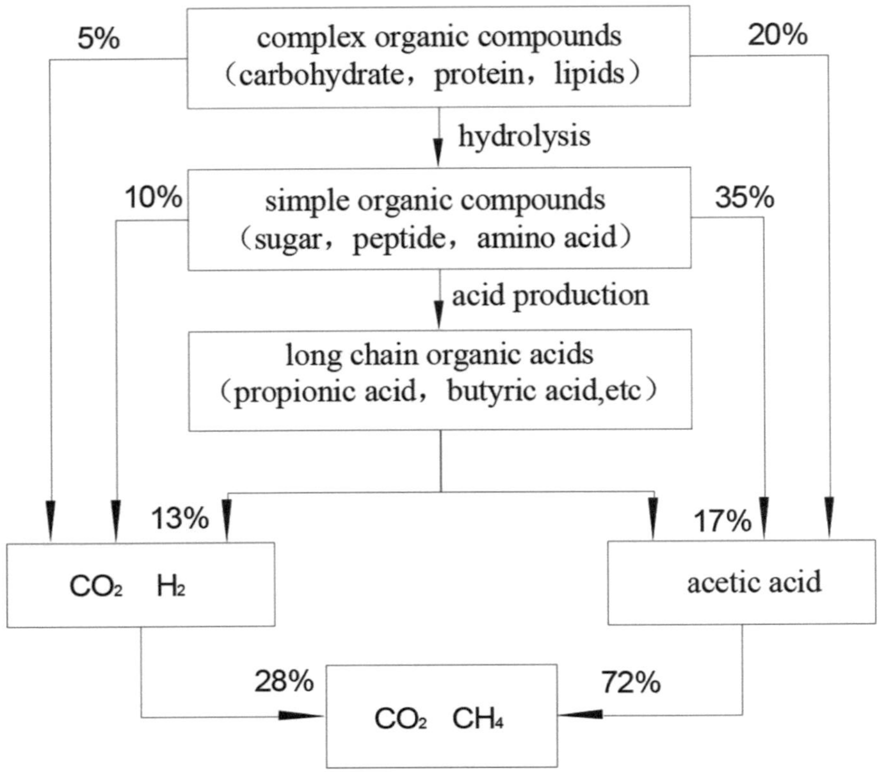

Figure 1.
Anaerobic biochemical reaction block diagram.

utilize the organic intermediates at the same rate as these intermediates are produced. Anaerobic biodegradation of organic matter can almost always be carried out spontaneously as soon as reaction conditions are suitable.

Anaerobic reactors have been used for anaerobic biodegradation of organic matter in water for a long time. An obvious problem recognized in early studies was that by simply increasing the amount of anaerobic sludge in an anaerobic reactor without good sludge-water mixing, the increase in reaction efficiency of the anaerobic reactor was limited. Subsequent studies have shown that using, for example, refluxed gas released from the anaerobic reaction or refluxed effluent to agitate the sludge-water and increase the strength of the sludge-water mixing also failed to significantly improve the reaction efficiency of the anaerobic reactor. It was not until Lettinga et al. developed the UASB anaerobic reactor in the 1970s and the development of the fixed-fill bed anaerobic reactor in the same generation. These types of anaerobic reactors have a higher anaerobic biochemical reaction efficiency than the previous anaerobic reactors that consisted of air bubbles or mechanical agitation of the sludge in suspension. The study of these anaerobic reactors and their engineering applications has gained much attention because of their well-shaped sludge morphology and the conditions conducive to sludge-water mixing. Speeces describes the special clustering of anaerobic microorganisms in granular sludge and biofilms as reducing the distance of metabolic material transfer and thus optimizing the collaboration between microorganisms [1]. It can be seen that in addition to the amount of anaerobic sludge and the state of sludge-water mixing, the morphology of the anaerobic sludge present, the stability of the anaerobic colonies within the sludge, and the solid phase mass transfer distance all influence the reaction efficiency of the anaerobic reactor. The morphology of the anaerobic sludge and the composition and stability of the internal biological community must be maintained while the sludge-water in the anaerobic reactor is adequately mixed. As a result, anaerobic reactors cannot use general mechanical or similar mixing methods to mix sludge-water simply by increasing the mixing intensity to avoid high shear flows destroying the anaerobic sludge morphology and the composition of its internal biological community. This makes sludge-water mixing in anaerobic reactors rather special and needs to be focused on both the structural design of the anaerobic reactor and the operating process, as well as the mixing pattern to mix sludge-water at low mixing intensities. This chapter will analyze ways to improve the efficiency of the sludge-water mixing reaction in terms of anaerobic reactor structure, operation, sludge properties, sludge-water mixing pattern, and mixing intensity. This will assist in the engineering design and operational control of anaerobic reactors to improve the efficiency of anaerobic biochemical degradation of organic matter.

1.2 Continuous stirred tank biochemical reactor (CSTBR)

According to the basic theory of mixed reactors, there are three mixing modes of reactors used for mixing reactants, mode of continuous stirring tank reaction (CSTR), mode of diffusion mixed reaction, and mode of push-flow reaction. Due to the long reaction time and complexity of the biochemical degradation of organic pollutants, the CSTR mode is more often used in biochemical reactors of wastewater treatment.

Figure 2 shows a model of an ideal CSTR. By definition, the reactants are instantaneously mixed in the reactor, with complete homogeneous mixing at time t = 0 and equal reactants concentrations at any point or a uniform concentration distribution in the reactor.

Let the reactor volume V, the input and output flow Q, the hydraulic retention time T, T = V/Q, the initial concentration of inert tracer C_0, and the output concentration C(t). In accordance with the mass balance equation, it is obtained that

$$\frac{C(t)}{C_0} = exp^{\left(-\frac{t}{T}\right)} \tag{1}$$

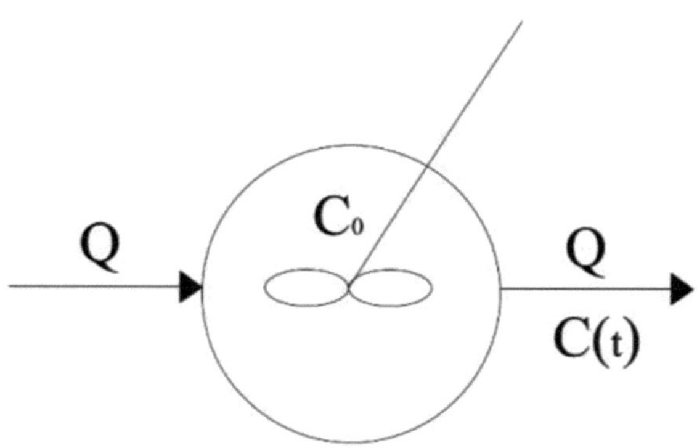

Figure 2.
Schematic diagram of CSTR reactor.

In practical engineering, regardless of the mixing method, the time $t > 0$ for homogeneous mixing in a reactor deviates to varying degrees from the ideal model in **Figure 2**. The concentration in a reactor can tend to be homogeneous, or the reactor tends to CSTR when $t \rightarrow \infty$. In designing a mixed reactor, the time t, when the mixing process tends to mix homogeneity, is much less than the reaction time T, $t << T$, the reactor can be approximated as a CSTR. As shown in **Figure 3**, increasing the mixing intensity of a mixing process with diffusion mode, its Peclet number decreases the output line E(t) of the tracer of instantaneous source tends to that of CSTR for a shorter time t. Therefore, suitable mixing methods can be selected or constructed according to the reaction time T and mixing time $\bar{t}$ in to save energy or meet other constraints required by the reactants. For example, anaerobic reactors require a low mixing intensity to complete the sludge-water mixing reaction.

By mixing the water through a turbulent flow field generated by high-intensity mechanical agitation, the reactor can approach CSTR indefinitely. Short reaction times and small volumes or mixing scales are easy for mixing with higher-intensity mechanical agitation to achieve CSTR. The biochemical degradation of organic matter has a slow reaction rate, takes longer to complete, and the volume and mixing space scale of the reactor is larger. If the biochemical reactor converges to CSTR at a time $\bar{t} << T$ relative to the biochemical reaction completion time or hydraulic retention time T, the mixing time $\bar{t}$ can be ignored, allowing the biochemical degradation of organic matter to proceed in a complete mixed state. The shorter the mixing time $\bar{t}$, the closer it is to CSTR, but at too short a mixing time $\bar{t}$ is unnecessary compared to the long biochemical reaction time T. For biochemical reaction tanks with large volumes and mixing space scale, the mixing mode should be constructed according to the time scale T of biochemical degradation of organic matter and mixing time scale $\bar{t}$, so as to obtain a biochemical reactor approximating to CSTR, $\bar{t} << T$. This type of reactor is defined as a Continuous Mixing Tank Biochemical Reactor (CSTBR), according to the sludge-water mixing time scale t and the biochemical reaction time scale T, $\bar{t} << T$.

Analysis of the large-scale turbulent mixing of the oxidation ditch and aeration basin can help to understand the mixing time scale $\bar{t}$ in a CSTBR. The aerobic-activated sludge allows the simultaneous mixing of sludge and water at higher turbulent intensities due to its high flocculation capacity. The mixing of small-scale turbulent eddies in the oxidation ditch and aeration tank tends to be instantaneous compared to the longer time taken to complete the biochemical reaction, so the

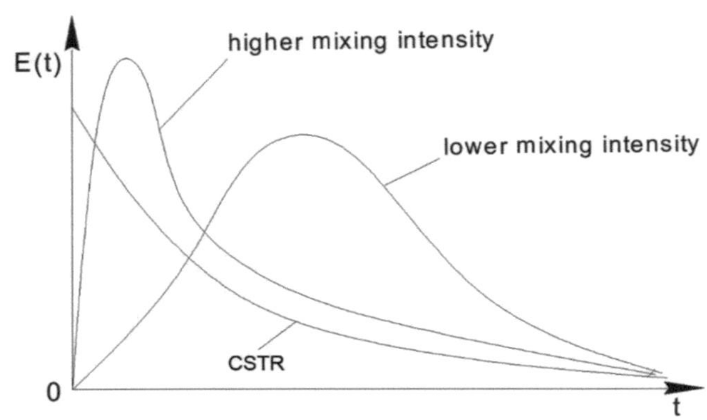

Figure 3.
E(t) curve of the actual complete mixing reactor.

mixing of small-scale turbulent eddies in the mixing process can be ignored, and only the mixing of large-scale time-averaged flows needs to be considered to analyze the mixing completion time t. Accordingly, Yang developed a dynamic mixing model of an oxidation ditch and an aeration tank to analyze the mixing processes and proved that the oxidation ditch tends to CSTR with a combination of two mixing modes: discontinuous mixing in circulation and one-dimensional continuous dispersing mixing, while the aeration tank tends to CSTR with discontinuous diffusion mixing [2]. Model calculations show that for the same biochemical reaction time T = 12.0 h, the mixing time $\bar{t}$ increases from $\bar{t}$ = 2.5 min for a small volume of V = 500.0 m^3 to $\bar{t}$ = 2.8 h for a large volume of V = 25,000 m^3 for oxidation ditches and aeration tanks with treatment of 1000–50,000 m^3/d and volumes of V = 500–25,000 m^3. **Figure 4** shows the model calculated mixing output E(t) curves for the instantaneous point sources of tracers in the oxidation ditch and aeration basin. As can be seen from **Figure 4**, the mixing reaction process in these two types of biochemical reactors is almost equivalent. In practice, the mixing is assumed to be complete in both the oxidation ditch and the aeration tank. The all-mixing completion time $\bar{t}$ = 2.5 min ~ 2.8 h satisfies the CSTBR condition, $\bar{t} << T$. Obviously, too short mixing completion time is unnecessary. For oxidation ditches with smaller volumes, the mixing time $\bar{t}$ can be increased by reducing the roughness of the ditch wall and the water flow velocity in the ditch; when the ditch volume is too large, the roughness of the ditch wall can be increased to shorten the mixing time $\bar{t}$, making $\bar{t} << T$. For aeration tanks, the aeration air rate can be adjusted to preset the mixing completion time $\bar{t}$.

Unlike in oxidation ditches and aeration tanks where sludge-water can be mixed at a high turbulent intensity. As anaerobic sludge lacks sufficiently strong flocculation capacity, anaerobic bacteria communities are easily destroyed by shear stresses in the flow. Thus, high-intensity mixing of sludge-water does not necessarily increase the efficiency of the anaerobic degradation reaction of organic matter. In the analysis of sludge-water mixing in anaerobic biochemical reactors, both the scale of the anaerobic sludge and the stability of the anaerobic bacteria community need to be considered, requiring greater control of the sludge-water mixing process.

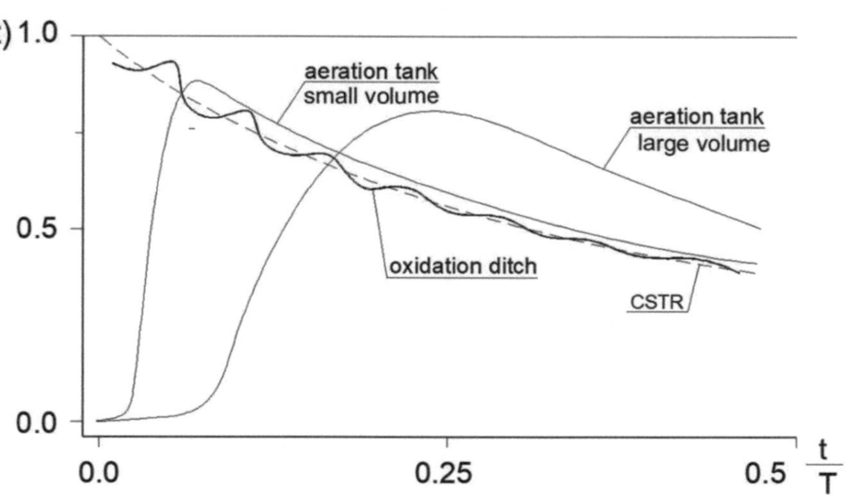

Figure 4.
Output pattern of oxidation ditch and aeration tank.

2. Methodology

2.1 Anaerobic sludge

Before discussing sludge-water mixing in anaerobic reactors, the properties of anaerobic sludge are analyzed. The main body of anaerobic sludge is made up of clustered aggregations of anaerobic microorganisms. These microorganisms adhere to each other synergistically, forming a highly structured and stratified symbiosis. For example, in anaerobic sludge, in the outer layer, it is the fermenting bacteria that are clearly dominant; while in the deep inner layer, the bacteria that degrade propionate are dominant. One important way to improve the biochemical reaction efficiency of an anaerobic reactor is to use large amount of sludge. However, it may also reduce the efficiency of sludge-water mixing, which is not conducive to anaerobic biochemical reactions.

The early anaerobic reactors were simple in construction. The reactor is generally a reaction tank with an inlet and an outlet and a gas collection port. The anaerobic reactor relies on the bubbles released from the anaerobic reaction to naturally mix the sludge and water, namely, biogas stirred anaerobic reactor. Sometimes, a mechanical stirring device is set up to disturb the deposited sludge and suspend it to participate in the sludge-water mixing reaction.

In the interior of these anaerobic reactors, the anaerobic sludge can migrate unrestrictedly within the reaction zone. In practice, anaerobic sludge of uncertain geometry and distribution can be clearly observed. The sludge can either float in larger blocks with geometries even larger than 100.0 mm or more, or it can break up into fine particles and remain suspended in the water. In addition, sludge can float on the water surface or be deposited on the bottom of the tank and collected in large sludge masses. The sludge-water mixing at any spatial location exhibits a stochastic character. This type of anaerobic reactor is defined as an unconfined sludge anaerobic reactor. The average agitation intensity of its bubbles is not high, and the sludge-water mixing reaction is inefficient and the time to complete the sludge-water mixing reaction is longer [3].

To date, no studies have been carried out on the strength of anaerobic sludge in relation to sludge-water mixing. Speeces's study noted that dense anaerobic sludge can be obtained in practice, but that the high intensity of turbulent shear flow can destroy the composition of bacterial aggregates in anaerobic sludge and that recovery can take weeks or months. The application of mechanical mixing of sludge-water requires a high mixing intensity just to keep the anaerobic sludge in suspension. Furthermore, apart from energy consumption considerations, the resulting complete sludge-water mixing does not necessarily increase the efficiency of the anaerobic reaction. Therefore, this mode of sludge-water mixing is rarely used in engineering now.

The geometry of the anaerobic sludge affects the efficiency of the anaerobic reaction. Larger sludge sizes can limit the anaerobic reaction rate due to the long transfer distance within the sludge, while smaller sludge sizes are not conducive to the formation of symbiotic biomass in the sludge, which is also detrimental to the anaerobic reaction. Using VFA, which is susceptible to anaerobic degradation, as an indicator, Speeces concluded that the unrestricted mass transfer distance for molecular diffusion within the anaerobic sludge is no greater than 1.0 mm and that at this scale, complete biopolymers are also obtained at the same time. For example, hydrogen-producing and hydrogen-using bacteria can both be in the same bacterial aggregate and the transfer of H_2 will not be a limiting factor in carrying out the methanation reaction. Speeces recommends an average sludge particle diameter of around 2.0 mm.

In UASB reactors and anaerobic reactors with fixed biofilms, the anaerobic sludge is confined to a local scale or fixed to the surface of the packed media and does not migrate with the flow. The geometrical scale of the anaerobic sludge can be controlled in the unrestricted mass transfer range. As in the UASB reactor, the anaerobic sludge particles are confined to the sludge bed with a sludge particle size of 2.0–4.0 mm, which is close to the unrestricted mass transfer distance within that of the Speeces recommended. This type of anaerobic reactor is defined as a confined-sludge anaerobic reactor [4]. With the appropriate sludge distribution, the confined sludge anaerobic reactors can achieve a complete sludge-water mixing reaction state through water mixing at a low mixing intensity, according to the condition of CSTBR.

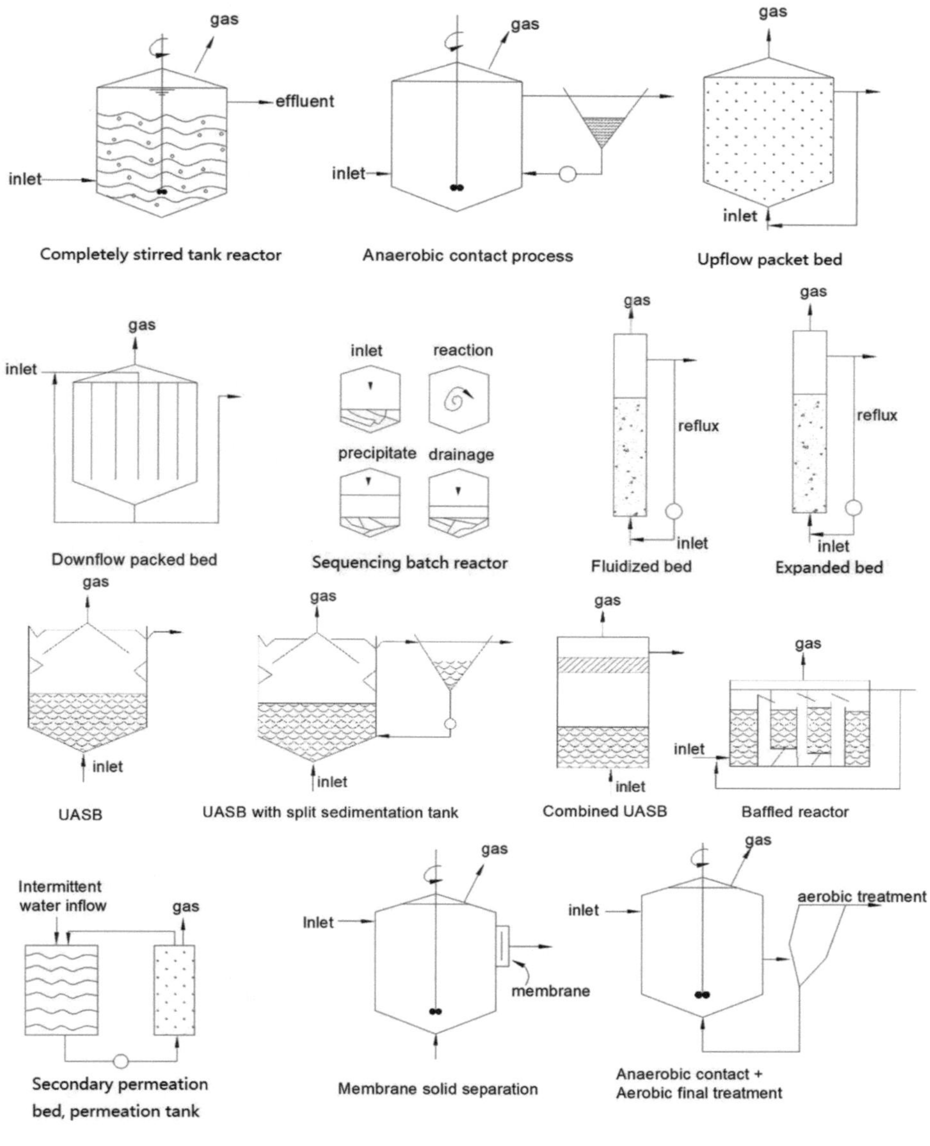

Figure 5.
The various types of anaerobic reactors.

2.2 Anaerobic reactor construction

The construction of the anaerobic reactor should be designed to achieve: (1) complete sludge-water mixing $\bar{t} << T$, operating in a CSTBR mode; (2) a low enough mixing intensity; and (3) a stable biological flora and unrestricted mass transfer scale in the solid phase sludge.

The configuration of the various types of anaerobic reactors currently in use is shown in **Figure 5**. The sludge bed of UASB reactor, packed bed, permeation in anaerobic reactor shown are all confined sludge areas, where the sludge is fixed or confined to a localized area, which is conducive to sludge-water mixing with low mixing intensity in CSTBR mode for presetting the sludge-water mixing state, by organizing the water flow pattern and shaping and distribution anaerobic sludge through the structural design.

In the anaerobic baffled reactor (ABR), although the anaerobic sludge is isolated in the reaction compartments and cannot migrate with the flow at the full reaction zone scale, it can migrate with the flow in the isolation compartments, and the sludge-water in each reaction compartment is mixed by bubble agitation. The ABR anaerobic reactor is not a confined sludge anaerobic reactor due to its large isolation compartments. The sludge moving space scale is of the same order of magnitude as the full mixing space scale.

Other anaerobic reactors, including fluidized bed and mechanically stirred anaerobic reactors are unconfined sludge anaerobic reactors, in which the anaerobic sludge moves within the reactor with the water flow. Complete sludge-water mixing reactions in fluidized beds and mechanically stirred anaerobic reactors require high flow velocity and mixing intensities. To date, there is still no effective method for analyzing or presetting the morphology of sludge and the state of sludge-water mixing in such anaerobic reactors. However, there are some lessons to be learned.

For anaerobic reactors with biogas-stirred, the construction is simple, and almost no internal structural design is required. Due to the uncertainty of the gas production of the anaerobic reaction, the anaerobic sludge-water mixing reaction is uncontrollable. All that can be achieved in current engineering applications is to apply a mixing model, which is posteriori, to analyze the state of sludge-water mixing and the volumes involved in the mixing reaction based on the monitoring of operational data.

3. Key results

3.1 UASB reactor

Figure 6 shows the internal structure of a UASB anaerobic reactor [5]. The main structure of which consists of five parts: (1) the sludge bed reaction area formed by the high-density accumulation of anaerobic sludge particles in the lower part of the reactor; (2) the sludge blanket; (3) the water distribution area at the bottom; (4) the three-phase separator; and (5) the gas collection chamber.

In **Figure 6**, the height of the sludge bed is H_1, and the volume is V_1; The sludge blanket with height H_2 and volume V_2, on the upper part of the sludge bed, below the three-phase separator; the sludge separating area above the inlet of the three-phase separator, with a height of H_3 and a volume of V_3.

The water enters the reactor from the bottom water distribution zone. After anaerobic biochemical degradation of the organic pollutants in the sludge bed reaction zone, it flows into the upper sludge blanket zone and out through the three-phase separator. All sludge particles overflow from the sludge bed with the rising water flow

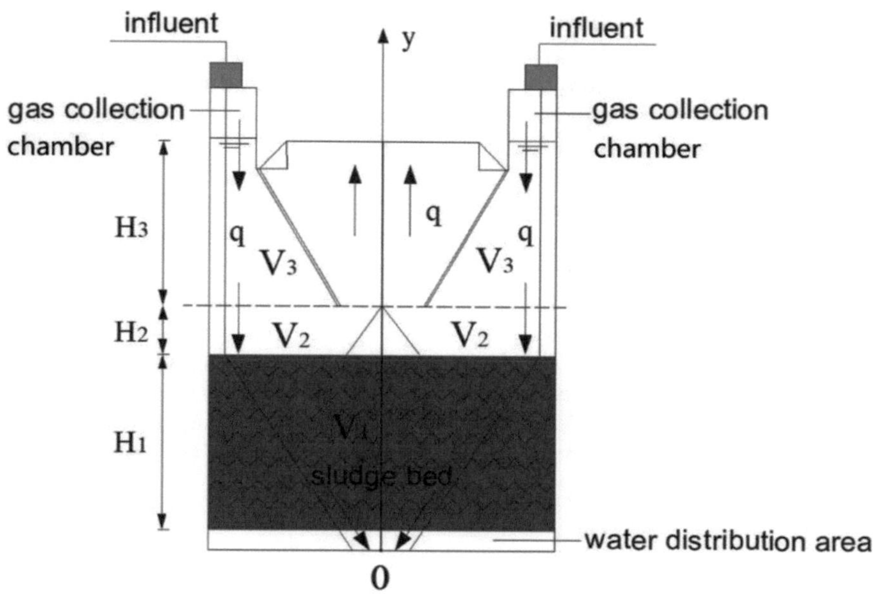

Figure 6.
Schematic diagram of UASB reactor.

and bubbles are separated in the V_2, V_3 zone, and the three-phase separator and settled back into the sludge bed, while the gases are collected in the gas collection chamber. The sludge bed is the main reaction zone where the organic matter is degraded. In the V_2 and V_3 zone, small amounts of suspended sludge particles are present, and the anaerobic biochemical degradation of organic matter continues, but in smaller quantities.

As can be seen from **Figure 6**, sludge-water mixing in the UASB reactor occurs in the sludge bed reaction zone V_1 and water in the upper zone ($V_2 + V_3$) with upwelling bubble agitation. The two mixing zones are adjacent to each other, but the mixing patterns are different. The maximum mixing volume that may participate in a UASB reactor is ($V_1 + V_2 + V_3$).

According to the study by Pol et al. [6], settling velocities of granular sludge of approximately 60 m/h are common, whereas the superficial upflow velocities in UASB reactors are usually kept below 2.0 m/h in practice. Therefore, the mixing in the sludge bed can be analogous to the diffusion of a seepage flow in general porous medium with dense sludge granular. The sludge blanket ($V_2 + V_3$) is stirred unstably by biogas at a small spatial scale, and the volume V_C involved in complete mixing cannot be preset，$V_C \leq (V_2 + V_3)$. Fortunately, the amount of suspended anaerobic sludge particles in this zone is small [7], and it is possible to disregard sludge-water mixing or ignore the amount of organic matter degraded and only analyze the mixing of the water.

Following the different sludge-water or water mixing patterns in these two mixing zones, mixing model equations can be developed, respectively [4]. Assuming that the anaerobic sludge particles in the sludge bed are distributed homogeneously, the sludge-water mixing process in the UASB can be simplified to a series-connected water advective diffusion unit V_1 with mixing unit Vc, where the water is stirred by biogas bubbles.

The water entering the sludge bed from the bottom is uniformly distributed and flows through the sludge bed in the form of seepage. The organic matter is biochemically degraded by the anaerobic sludge while the water diffusive mixing in the sludge bed, and the two processes interact with each other. However, If the anaerobic biochemical degradation reaction in the sludge bed operates in a CSTBR mode when the complete time scale t_P of the diffusive mixing is sufficiently small, $t_P << T$, or at the time t, $t < t_P$, before mixing homogenization, the influence of biochemical degradation of organic matter can be disregarded, and only inert tracers are required to analyze the diffusive mixing process of the water. For anaerobic sludge particles of homogeneous distribution, the mixing of sludge-water and mixing of water is completed simultaneously in the sludge bed.

As shown in **Figure 6**, $u = q/A$, where u is the velocity of seepage flow, q is the influent, and A is the section area. Let C (y, t) be the tracer concentration distribution in sludge bed, $y \in (0, H_1)$. Analogous to the diffusion of a seepage flow, the one-dimensional concentration equation of the tracer in the sludge bed is as follows:

$$\frac{\partial C(y,t)}{\partial t} + u\frac{\partial C(y,t)}{\partial y} = E_y\frac{\partial^2 C(y,t)}{\partial y^2} \tag{2}$$

where E_y - diffusion coefficient of the sludge bed.

The sludge blanket ($V_2 + V_3$) is stirred unstably by biogas, and the volume involved in complete mixing cannot be preset. V_C is the volume involved in complete mixing, $V_C \leq (V_2 + V_3)$ and letting C (t) be the tracer concentration in V_C, then the tracer concentration equation of V_C can be written

$$\frac{dC(t)}{dt} = \frac{V_c}{q}(C(H_1,t) - C(t)) \tag{3}$$

where $C(H_1, t)$ - tracer concentration of sludge bed at $y = H_1$ and time t.

Applying Eqs. (3) and (4) to simulate the diffusive transport process of the instantaneous tracer at the location of the inlet cross-section, the E(t) output curve, and the concentration distribution of the tracer, the time t_P for the completion of sludge-water mixing in the UASB reactor can be obtained to analyze the state of sludge-water mixing.

Let the tracer mass M be injected instantaneously at the inlet section. The initial boundary conditions of Eq. (3) of the sludge bed are.

$$C\,(0,0) = M\delta(y) \tag{4}$$

where $\delta(y)$ is a Dirac function.

$$C(y,0) = 0, y \in (0, H_1) \tag{5}$$

$$\frac{\partial C(0,t)}{\partial y} = 0 \tag{6}$$

The initial boundary conditions of Eq. (4) are.

$$C(0) = 0 \tag{7}$$

$$C(t) = C(H_1, t) \tag{8}$$

If V_C is known, based on the model Eqs. (2) and (3), and the initial boundary conditions (4) and (6), the E(t) output curve of the UASB inlet instantaneous tracer injection can be numerically calculated using the difference method.

Figure 7 shows the vertical distribution of tracer concentrations in the UASB reactor at moments t_i, i = 1, 2,...,10, obtained from the model analysis. As the mixing time increases, the peak tracer concentration gradually decreases and the sludge-water mixing area in the sludge bed becomes homogeneous. The concentration in the upper bubble natural mixing zone is shown at the end of the fold on the left side of the curve. Before the sludge-water in the sludge bed is uniformly mixed, the tracer concentration in the bubble agitation zone is lower than the tracer concentration in the outflow from the top of the sludge bed, and the concentration on the boundary surface at the top of the sludge bed decreases transversely. After homogeneous mixing, the tracer concentration in the bubble agitation zone is greater than that in the outflow from the top of the sludge bed, and the concentration variation in the two zones on the decomposition surface at the top of the sludge bed tends to be smooth. According to Yan and Yang's model, the sludge-water mixing in the sludge bed is considered complete when the peak concentration $C = 2\overline{C}$, $\overline{C}$—the average concentration of tracer in the sludge bed. In **Figure 7**, the time of complete mud-water mixing is t4 and the peak concentration of tracer is located in the central region of the sludge bed; when $t_4 << T_1$, the sludge bed operates as a stand-alone CSTBR, T_1—the hydraulic residence time of the sludge bed. And the upper bubble mixing region is the CSTR for the water mixing with a mixing volume of V_C.

Figure 8 shows the $E(H_1, t)$ output curves for the tracer at the top of the sludge bed and the E(t) output curves for the tracer in the upper bubble mixing zone obtained from the UASB model calculations and analysis. The left side shows the high sludge bed expansion condition and the right side shows the low sludge bed expansion condition. In the high expansion sludge bed condition, the volume of the upper bubble mixing zone ($V_2 + V_3$) and the volume of water involved in mixing are small, $E(H_1, t)$ is close to the E(t) output curve, and the UASB reactor can be simplified to a CSTBR

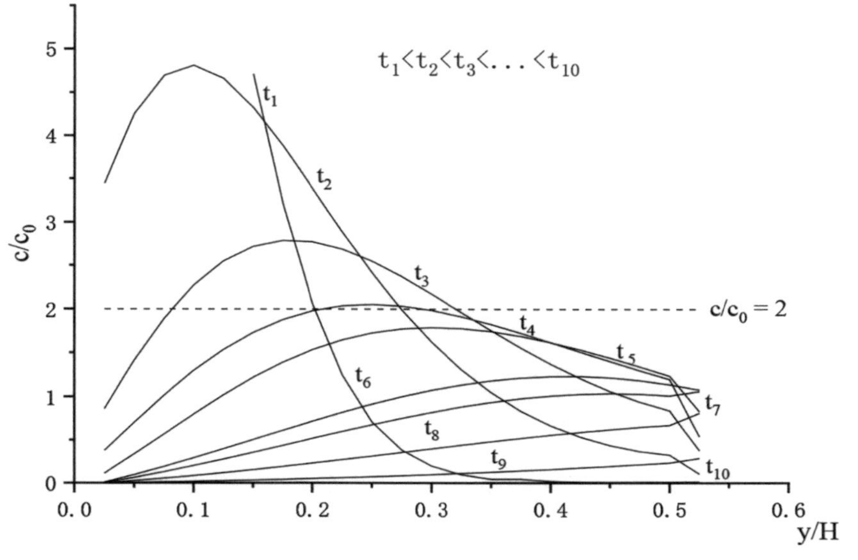

Figure 7.
Vertical concentration distribution curve of tracer.

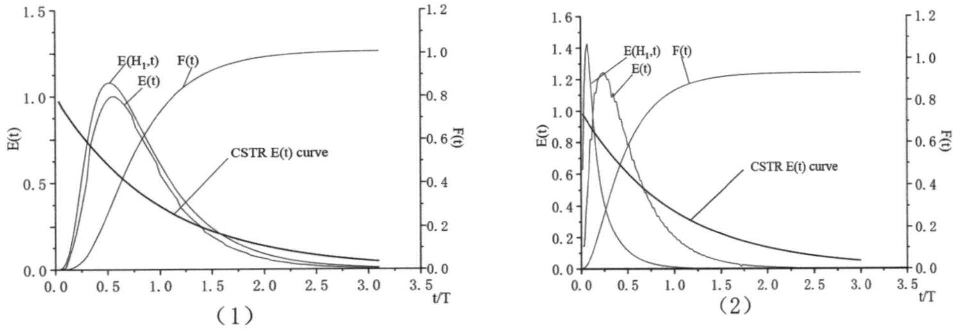

Figure 8.
UASB output concentration and cumulative mass output curve.

operating unit; in the low expansion sludge bed condition, $E(H_1, t)$ deviates from the $E(t)$ output curve, and the UASB reactor can be simplified to a CSTBR of sludge bed operates in tandem with a CSTR of water with a mixing volume of V_C in its upper part. The tracer concentration peaks at high sludge bed swelling conditions are smaller than those at low sludge bed swelling conditions, and the sludge-water mixing is better.

In a computational analysis of the sludge-water mixing process in a group of UASB reactors, Yan and Yang used the UASB tracer test data from the Pena test on engineering scale to determine the diffusion coefficient of the sludge bed, $E_y = 0.00006 \sim 0.00012\ m^2/h$. The value of E_y is larger when the upward water flow velocity is larger, and the sludge bed expands and is smaller when the upward water flow velocity is smaller. The variation of E_y is not significant and is stable at 10^{-4} orders of magnitude.

Here present the key elements of a computational analysis of sludge-water mixing in a UASB reactor [2]. The sludge bed height $H_1 = 1.0 \sim 2.5$ m, an upward flow velocity $u = 0.4 \sim 0.8$ m/h, a hydraulic residence time $T_1 = 3.1 \sim 3.8$ h, and a mixing completion time $t_4 = 1.0 \sim 1.9$ h, $t_4 << T_1$.The sludge bed operates as a CSTBR, with low upward flow velocity u and mixing intensity.

As V_C is generated by the agitation of bubbles released from the anaerobic reaction, its value cannot actually be determined in advance. However, there are two extremes in the operating conditions of the UASB reactor. The sludge-water mixing process can be calculated without prior determination of the V_C. This can be used to analyze sludge-water mixing characteristics of UASB reactor.

1. Sludge bed with high expansion condition, $H_2 = 0$, $(V_2 + V_3)$ is much smaller than V1, at this time can ignore the upper involved in the volume of water mixing V_C, calculate and analyze the state of sludge-water mixing in the sludge bed. Under these conditions, the sludge bed has the longest diffusion mixing time and flow, the percolation diffusion coefficient is also larger, and the sludge-water mixing state is the best.

2. The sludge bed with low expansion condition, where H_2 and $(V_2 + V_3)$ are maximum, but taking $V_C = 0$, $(V_2 + V_3)$ operates in a push-flow mode. At this condition, the sludge bed has the shortest seepage diffusion mixing time and flow and also has a smaller percolation diffusion coefficient due to the denser sludge particle build-up, which results in the worst mud-water mixing conditions

in the sludge bed. When (V_2 + V_3) is operated in push-flow mode, the end of the E (H_1, t) curve at the top of the sludge bed does not produce a descending fold when the boundary concentration of the sludge bed outflow is higher. As can be seen from **Figure 7**, the sludge-water mixing time calculated for this condition will be greater than the mixing time for the normal condition of $V_C > 0$. This condition occurs at the beginning of the UASB start-up operation when gas production is low.

From these two extreme operating conditions, the possible sludge-water mixing states in the UASB reactor can be approximated, while the other general operating conditions have sludge-water mixing states in between. In general operation, $0 < V_C < (V_2 + V_3)$. To obtain an accurate V_C during UASB reactor operation, a tracer test is required, and V_C can be calculated by model analysis of E(t) curve of tracer output.

The model analysis determines the seepage diffusive mixing of the sludge bed and the mixing of the upper water volume to complete the vertical mixing of sludge-water in the UASB reactor in CSTBR mode. The horizontal sludge-water mixing operation is ensured by a uniform water distribution system at the bottom of the sludge bed, resulting in a uniform horizontal concentration distribution in each section of the UASB. The uniformity of water distribution at the bottom of the sludge bed directly impacts the sludge-water mixing in the sludge bed. There is already more technical and engineering experience in this aspect. A more reliable water distribution system for UASB reactors is the small resistance distribution system. Although the accuracy of the water distribution uniformity is not as good as with large resistance distribution systems, its large cross-sectional flow paths make them less likely to block and easier to clear.

The morphological shaping of sludge particles and their scale control also are key factors in the control of sludge-water mixing. The shape and size of the sludge particles, as well as the compactness and homogeneity of the accumulation, have a direct impact on the seepage and diffusion processes in the sludge bed. Numerous biological factors—may influence the formation of sludge particles Speece [1], in addition to the usual COD concentration: alkalinity, organic acids, hydrogen partial pressure, multiple trace elements, methanogen types, lipids, and nitrogen and calcium. The formation of sludge particles appears to be a very complex biological process. However, once the sludge particles in the sludge bed have been initially formed, they can be hydraulically graded to leave the sludge particles with good settling and conformational scales. This results in a uniformly stacked sludge bed with the right size of anaerobic sludge particles. A range of 9.0–55.0 m/h is the recommended flow velocity for hydraulic classification by Speece [1].

The larger sludge bed expansion height H_1 and lower upward flow velocity u facilitate sludge-water mixing, allowing the UASB to operate in a CSTBR mode. However, a higher sludge bed expansion height will not be conducive to maintaining a uniform sludge particle distribution and releasing air bubbles between particles inside the sludge bed, it will also cause the sludge-water mixing reaction to deviate from CSTBR. Too low a flow velocity may result in compaction of the sludge bed, which is not conducive to seepage diffusive mixing. As a general rule of thumb, UASB reactors have an upward flow velocity of u = 1.0–5.0 m/h and a sludge bed height of H_1 = 1.5–3.0 m. Limits the height of the sludge bed and the upward flow rate, which also limits the amount of sludge and the volumetric load of organic matter in the UASB reactor.

3.2 Internal reflux packed-bed anaerobic reactor (IRPAR)

Internal reflux is widely used in biochemical treatment processes to promote mixing of the water and improve the efficiency of the biochemical reactions. In many cases, internal reflux is provided to facilitate the mixing of water, although it may be mainly required to meet other process operating conditions. For example, in aerobic or anaerobic biofilters, internal reflux is used to increase the flow velocity to flush out excess or aged biofilm growing on the surface of the filter media; in fluidized beds, internal reflux is used to increase the water flow velocity to fluidize suspended bioparticles and also to mix the water; in filled bed anaerobic reactors, internal reflux is used to increase the mixing intensity of bubbling between the pores of the filler and the mixing of the water and is desired to obtain a completely mixed anaerobic reactor.

In **Figure 9**, the anaerobic reactor IRPAR is equipped with an internal reflux system [7]. The substrate was fed in at the bottom of the reactor through a conical inlet followed by a distribution plate. The effluent was withdrawn from the top for recirculation and disposal.

Here the water and sludge-water mixing processes in an IRPAR are analyzed, and the convergence to a CSTR biochemical reactor as a CSTBR is demonstrated under set conditions. This analysis and demonstration were done by Yan and Yang.

Figure 9 shows the model of an IRPAR. C_0 is the influent concentration, and C(H, t)is the effluent concentration at time t. The height and the volume of the packed bed reaction zone are H and V, and the reflux ratio is R, $R >> 1$. The influent q and the reflux Q, $Q = Rq$, are mixed instantaneously at Point A and uniformly distributed to

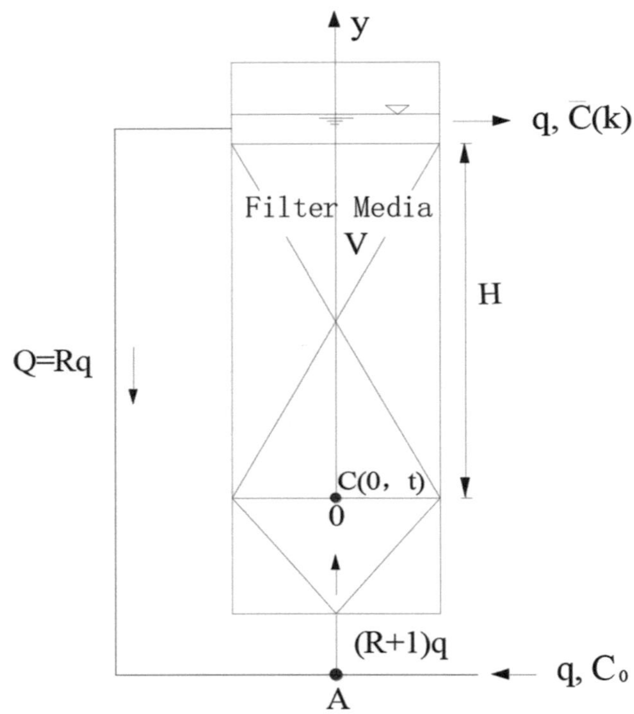

Figure 9.
Model of IRPAR.

the bottom with concentration $C(0, t)$. The hydraulic retention time is T, $T = V/q$, and the reflux cycle time is $T_L = V/Q$.

For the fillers with pore sizes much smaller than the mixing scale of the anaerobic reaction zone, the sludge-water mixing proceeds with reflux water and influent traversing the surface of the biomembrane in each cycle. As shown in **Figure 9**, the mixing of water occurs at three scales of mixing: (1) large-scale transport of water flowing from the upper outflow to the bottom inflow, and then traversing the surface of the biomembrane attached to the filler in the reactor; (2) small-scale mixing of the biogas and the rising water; and (3) the microscale molecular diffusion within the biomembrane on the filler. The large spatial and time scales of the reflux transport are close to H and T_L, respectively. The small mixing scale of the rising flow and biogas is close to the scale of the filler pores. The mixing scale within the biomembrane is close to the membrane thickness, which is much smaller.

Figure 9 ignores the flow time of the internal circulation pipeline. Let C (y, t) be the concentration distribution of the tracer, $y \in (0, H)$; $\overline{C}$ (k) is the average tracer concentration in the reactor at time $t = kT_L$, k = 0, 1, 2, . . ., Eq. (3) was derived:

$$\overline{C}(k) = \frac{1}{T_L}\int_{kT_L}^{(k+1)T_L} C(H, t)dt \tag{9}$$

According to Eq. (6), the average output concentration in T_L is equal to the average concentration $\overline{C}$ (k) in the reactor. The average concentration $\overline{C}$ (k), k = 0, 1, 2, ..., can be obtained by mathematical induction

$$\frac{\overline{C}(k)}{\overline{C}(0)} = \left(1 - \frac{1}{R}\right)^k \tag{10}$$

where $\overline{C}$ (0) - initial average concentration in the reactor and k = 0. According to Eq. (8), furthermore, as $R \to \infty$, the average concentration $\overline{C}$ (k) achieved in a CSTR at $t = kT_L$, can be obtained as follows:

$$\frac{\overline{C}(k)}{\overline{C}(0)} = e^{-\frac{k}{R}} \tag{11}$$

The cycle time T_L can be defined as a mixing time scale of the IRPAR. The essence of internal reflux mixing is large-scale transport, and it cannot change the tracer concentration distribution C (y, t). Only the small-scale mixing of biogas and the rising water flow can encourage uniform mixing of the water.

The discontinuous discrete distribution of the mean tracer transport concentration on the mixing time scale T_L is obtained from Eq. (10) and shown in **Figure 10**. The CSTR continuous output curve is also given in the figure. From **Figure 10**, it can be seen that after $R > 10$, the concentration output curve of the anaerobic reactor has been able to discretely converge to the continuous concentration curve of the CSTR as a CSTBR. A simple derivation shows that the rising flow velocity in the reactor with internal reflux is $u_L = (R + 1)u$, u being the rising flow velocity in the absence of reflux or when R = 0. For u = 0.5 ~ 5.0 m/h, and taking R = 20, u_L = 10.5 ~ 105 m/h, $T_L/T = 21$. At this point, the mixing time scale $T_L << T$, and the ability to mix sludge-water at low flow velocity and mixing intensities is close to CSTR on the time scale T_L, as a CSTBR. Taking a larger R-value and reducing T_L can be closer to CSTR or be a

CSTBR, but will make u_L too large for the stability of the microbial symbionts in the anaerobic sludge membrane.

Consider the steady-state anaerobic degradation reaction of organic pollutants with only internal reflux transport. The concentration difference ΔC of organic pollutants formed between the bottom section and the upper outlet section is

$$\triangle C = \frac{C_0 - C(0,t)}{R} \quad (12)$$

From Eq. (12), letting $R \to \infty$ or $T_L \to 0$, Eq. (13) was obtained

$$C_0 + Tr = C(H,t) \quad (13)$$

where r is biochemical degradation rate of pollutants.

Eq. (13) is the concentration equation of a steady-state CSTR. Therefore, considering the homogeneous distribution of the anaerobic biomembrane, even without considering the unstable small-scale mixing of biogas and rising water flow, and only using a sufficiently large internal reflux ratio R, $R >> 1$, an equivalence between the sludge-water mixing reaction in the internal reflux anaerobic reactor and the complete mixing and reaction of the CSTR can be achieved, and the microscale molecular diffusion of the biomembrane is processed synchronously.

In accordance with the conditions set by the internal reflux sludge-water mixing model, the pore size of the filler media and the pore space between the filler media in a packed bed anaerobic reactor should be sufficiently small, much smaller than the mixing space size of the anaerobic reactor, and evenly distributed. In addition to economic considerations, the specific surface area, roughness, biological inertia, mechanical strength, and suitable shape and void space of the filler material should be considered. The particle size and inter-particle void size of the filler is a key consideration, with Speece recommending a particle size of around 15 mm. Speece's research shows that the depth of the packed bed has no significant effect on operation, but the recommended packing height is 2.0 m [1]. Although the smaller filled bed height

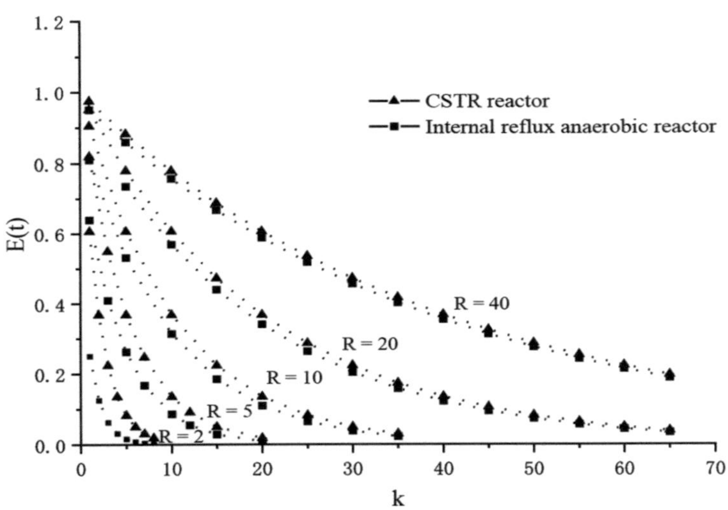

Figure 10.
E(t) curve of the CSTR and the internal reflux anaerobic reactor.

increases the area of lateral water distribution and the difficulty of uniform water distribution, smaller u_L, and mixing intensity can be obtained.

Filled bed anaerobic reactors rely on the mixing of water at point A in **Figure 9** and uniform water distribution at the bottom of the packed bed to ensure a uniform lateral distribution of concentration in the packed bed. Similar to the UASB reactor, the water distribution at the bottom of the packed bed in a packed bed anaerobic reactor uses a small resistance distribution system to help prevent blockages in the water distribution system and to facilitate cleaning and maintenance.

In some aerobic or anaerobic biochemical reactors with filled media and internal reflux, when the particle size of the filled media is much small and evenly distributed, the sludge-water mixing process in large-scale internal reflux is isomorphic. A large-scale sludge-water mixing model similar to that of the IRPAR can be used to analyze the operating mode of sludge-water mixing in internal reflux and the conditions for achieving a CSTBR, such as biofilters and biological contact aerobic or anaerobic biochemical reactors.

4. Conclusion

According to anaerobic sludge properties, sludge-water mixing in anaerobic reactors operating in complete mixing mode should all take place at low mixing intensities or water flow velocity in order to maintain a stable anaerobic biota in the anaerobic sludge and thus improve the efficiency of the anaerobic biochemical reaction. In this chapter, the CSTBR was defined in terms of the sludge-water mixing time scale $\bar{t}$ and the biochemical reaction time scale T ($\bar{t} << T$), which is used to analyze the sludge-water mixing characteristics of anaerobic reactors and tends to CSTR as $\bar{t} \rightarrow 0$. Both the UASB reactor and the IAPRA reactor are confined sludge anaerobic reactors and have different mixing patterns. However, both can achieve CSTBR mode with low mixing intensity and simultaneous sludge-water mixing. According to the structural characteristics and mixing patterns of these two types of anaerobic reactors, corresponding sludge-water mixing calculation models can be established to analyze the working conditions and corresponding operational parameters to achieve the CSTBR mode.

1. The UASB reactor consists of a sludge-water mixing reaction unit that operates as a CSTBR and a unit of water mixing unit that operates as a CSTR in series mode. The UASB reactor can be simplified to a single-stage CSTRB anaerobic reactor when the sludge bed expansion height is large and the volume of the upper part of the water involved in the natural mixing of the bubbles is small. The sludge-water mixing reaction can reach the CSTBR at a low mixing intensity, which corresponds to flow velocity as low as 0.8 m/h in a UASB of engineering scale.

2. IAPRA mixes sludge-water with a internal reflux of large space scale, and the mixing time scale is the internal reflux period T_L. The sludge-water mixing reaction reaches CSTBR when $T_L << T$. The sludge-water mixing intensity is lower, and the corresponding water flow velocity can be as low as below 5.0 m/h for a CSTBR mode.

Author details

Yang Min
Faculty of Geosciences and Environmental Engineering, Southwest Jiaotong University, Chengdu, Sichuan, China

*Address all correspondence to: yang_minxna@163.com

References

[1] Speece RE. Anaerobic biotechnology for industrial wastewater. Translated by Yaxin Li. Beijing: China Architecture and Building Press; 2001

[2] Yang M. Dynamic mixing models and analysis of the mixing processes for an oxidation ditch and an aeration tank. Water Environment Research. 2022;**94**(6):e10742

[3] Yang M, Wu P, He W, He J, Li W. Study on start-up characteristics of anaerobic reaction of high VFA concentration pomace waste liquid. Environmental Engineering (China). 2015;**33**(8):32-36

[4] Heng Y, Min Y. A Sludge-Water Mixing Model of Anaerobic Reactors with Confined Sludge. Journal of Environmental Engineering. 2022;**148**(11):04022066

[5] Peña M, Mara D, Avella G. Dispersion and treatment performance analysis of an UASB reactor under different hydraulic loading rates. Water Research. 2006;**40**(3):445-452

[6] Hulshoff Pol LW, de Castro Lopes SI, Lettinga G, Lens PNL. Anaerobic sludge granulation. Water Research. 2004;**38**(6):1376-1389

[7] Liu BY, Pfeffer JT, Suidan MT. Equilibrium model of anaerobic reactors. Journal of Environmental Engineering. 1995;**121**(1):58-65

Chapter 3

The Investigation of Chemical Composition and the Specific Heat Capacity of Cow Dung and Water Mixture

Vhutshilo Nekhubvi

Abstract

Energy is essential for the progress and development of nations. It must be reliable, affordable, and environmentally friendly. Among the most promising renewable energy sources, biogas technology has been developed to secure the existing energy supply. However, there is a need for more scientific research on the optimal use and performance of biogas plants for beneficiaries and installers. This study investigated the chemical composition of cow dung and its specific heat capacity. The results show that elements such as Al_2O_3, CaO, Fe_2O_3, K_2O, MgO, MnO, Na_2O, P_2O_5, SiO_2, and TiO_2 have different chemical compositions. Furthermore, the results show that cow dung's composition and oxide content affect its specific heat capacity. Dzwerani had the highest concentrations of Al_2O_3, Fe_2O_3, and SiO_2. Since the oxide composition of the dung samples from Tshino, Maila, and Gogogo differed, their specific heat capacities were also different. The results of this study encourage further investigations to determine a more accurate relationship between specific heat capacity and oxide composition.

Keywords: biogas, cow dung, specific heat capacity, titanium dioxide, silicon dioxide

1. Introduction

As nations strive to progress, they need energy to power their industries, fuel transportation, and provide electricity for homes and businesses. Additionally, access to energy also allows countries to pursue other goals, such as improving healthcare, education, and infrastructure. Without reliable access to energy, economic growth is severely limited, as many industries and businesses require energy to operate. Furthermore, access to energy can also provide a source of employment to citizens, creating jobs, and helping to lift people out of poverty. As a result, almost all developing countries have taken initiatives to introduce biogas technology in rural areas to improve energy supply and reduce poverty [1–5]. However, the beneficiaries and installers of these plants still need more scientific knowledge about biogas production. This means that they need help to make the best use of the technology available and optimize the performance of their plants. Due to its availability, cow dung is the most common feedstock used in household biogas digesters in rural areas. The investigation results of the study [6] showed that cow dung might be one of the feedstocks for efficient biogas production. Fresh cow dung is estimated to contain 28% water [7]. However, Refs. [8, 9] indicated that fresh cow dung contains approximately 80% water. For anaerobic digestion (AD) to generate biogas energy, fresh cow dung is mixed with water at a widely-used ratio of 1:1 [6]. Cow dung moisture is an important parameter that influences biogas production. In the absence of moisture, the anaerobic bacteria responsible for biogas production cannot function properly, and the process slows down. If moisture levels are too high, sludge may form, clogging the digester and reducing efficiency. Research shows that cow dung slurry comprises 1.8–2.4% N_2, 1.0–1.2% P_2O_5, 0.6–0.8% potassium, and 50–75% organic humus [10]. The main issue for biogas energy production lies in knowing several parameters in the mixture, building materials of the digester, insulation, and a heat source. Alfa et al. [11] indicated that biogas production depends on the physical and chemical properties of the feedstock type used. Cow dung's physiochemical properties are important in operating a biogas digester system and maintaining digestion stability [12, 13].

Temperature is one of the most important parameters of AD system [14]. It was shown that the temperature of the AD system's manure directly influences biogas production. They found that the higher the temperature, the higher the biogas production [15]. This indicates that thermal energy management is crucial for the efficiency of anaerobic digestion. Obot et al. [16] indicated that heat transfer problems are associated with thermal properties, one of which is specific heat capacity. Obot et al. [16] argued that specific heat capacity is critical for thermal analysis problems. Specific heat capacity is defined as the measure of the amount of heat energy required to raise the temperature of a slurry inside the digester by 1°C [16]. The heat demand of AD systems depends on substrate characteristics, operating temperature, geographic region, and AD parameters such as digester type or size [17]. Thus, knowing the specific heat capacity of cow dung and the water mixture added to the digester is beneficial when determining the amount of heat required to raise the slurry temperature to the desired operating temperature. This information can help select the appropriate heating system and equipment for a biogas digester, leading to an efficient and effective biogas production process. As a result, the biogas production process is optimized, leading to a more cost-effective and reliable energy source. Methods for determining the specific heat capacity of material have been reported. However, few reported specific heat capacity of cow dung and water mixture. Gebremedhin et al. [18] assumed that the specific heat of cow dung is equal to the specific heat of water when modeling heat transfer problems in their study. Nayyeri et al. [19] conducted a study aiming to determine the thermal properties of cow dung. Only three physical quantities were determined: specific heat capacity, thermal conductivity, and thermal diffusivity. They reported that the specific heat capacity of cow dung increased linearly from 1.992 to 3.606 kJ/kg°C. Yerima et al. [20] reported the specific heat capacity of cow dung to be 2.0525kJ/kg°C. Ref. [21] reported the specific heat of cow dung to be 2.7992 kJ/kg°C and further showed that the specific heat capacity of the slurry mixture of water and cow dung is the sum of the specific heat of the water and that of cow dung (3.490 kJ/kg°C). Das and Mondal [22] used the following expression for the specific heat capacity of the slurry of cow dung and water mixture added to the digester when determining the amount of heat required to raise the slurry temperature to the digester operating temperature of 35°C

$$C_p = 4.19 - 0.00275(\mathrm{TS}) \tag{1}$$

where ?? is the specific heat capacity$\left(\frac{\mathrm{J}}{\mathrm{kg}}\,^\circ\mathrm{C}\right)$ of the slurry, and (TS) is the total solid of cow dung expressed in $(\mathrm{kg/m^3})$.

This research aims to investigate the chemical composition and specific heat capacity of cow manure-water mixtures, with a special focus on their suitability for biogas production. Anaerobic digestion can be optimized by investigating the thermal properties of cow manure-water mixtures, which increase the biogas production efficiency. Different cow manure-water mixtures will be tested under different conditions, including temperature and mixing ratio, to measure their specific heat capacities. Additionally, we will analyze the chemical composition of cow dung to identify the main components that contribute to biogas production. Research results from this study will contribute to a better understanding of cow dung's potential as a biogas feedstock. To improve the efficiency and sustainability of biogas production from cow dung-water mixtures, we must understand the digester's chemical composition and specific heat capacity.

2. Methods

2.1 Analytical methods of chemical composition

After arrival from the feedlot, the quantity of fresh cow dung samples was heated in an oven to 103°C for 24 hours to remove the moisture content. The procedure was carried out following the Standard Methods for the Examination of Water and Wastewater [23]. The samples were milled with the vibrating disc mill RS 200 until the final analytical fineness (< 40 μm) was reached. Each 10 g of milled powder was mixed with boric acid before palletization. The powder was then pressed onto a bed of 2.5 g boric acid in an aluminum cup (40 mm diameter), applying a pressure of 30 tonnes for 20 seconds. The chemical composition was measured using S2 Ranger (Bruker) X-ray fluorescence (XRF) in a vacuum with a 30 mm sample holding mask. The pellets were introduced into the XRF instrument after calibration. The measurement results were displayed as main oxides (%) or trace elements (ppm).

2.2 Determination of specific heat capacity

A recent study found that there is no universal method for determining specific heat capacity [24]. An experimental method is described for determining the specific heat capacity of cow dung mixed with water in the study. Researchers used the method of mixtures to determine the specific heat capacity of a mixture, which involves the combination of two substances with known properties. The method consists of the mixture of cow dung sample and water of known mass ($m_s = 0.1$ kg) and temperature $T_i(^\circ\mathrm{C})$ placed in an aluminum calorimeter mass,$m_{Al} = 0.033$ kg specific heat capacity $C_{\mathrm{Al}} = 0.900\ \mathrm{J/g^\circ C}$. A brass metal of mass $m_b = 0.20$ kg and specific heat capacity ($C_b = 0.38\ \mathrm{J/g^\circ C}$) was then heated to the desired temperature of 65°C using a water bath and immediately dropped into the mixture before wiping excess water. The equilibrium temperature was recorded as $T_e(^\circ\mathrm{C})$. The method determines the specific heat capacity by cooling the hot brass sample in cool cow dung and equating the heat losses of the brass metal with the heat gains of the cool cow dung slurry [19, 25]. Mathematically, the equation is written as:

$$C_s = \frac{m_b C_b (T_b - T_e) - m_{Al} C_{Al} (T_{Al} - T_e)}{m_s (T_e - T_i)} \quad (2)$$

As a result of this method, it is possible to determine the specific heat capacity of cow-dung-water mixtures by comparing the heat losses and gains during heat transfer.

3. Results and discussion

3.1 Chemical composition and specific heat of cow dung

According to the present results in **Table 1**, four different sites (Dzwerani, Tshino, Maila, and Gogogo) have different chemical compositions based on various elements such as Al_2O_3, CaO, Fe_2O_3, K_2O, MgO, MnO, Na_2O, P_2O_5, SiO_2, and TiO_2. It has been demonstrated that these results are relevant to biogas production using organic materials found at these sites. Several factors, including the chemical composition of the organic material, temperature, pH, and the presence of inhibitors influence biogas production efficiency. Biogas production is highly dependent on the chemical composition of the organic material. It is imperative to note that organic materials' chemical compositions can vary from place to place. Consequently, determining the feasibility of biogas production requires knowing the chemical composition of organic materials at different locations. According to the results, site A has a high SiO_2 and Al_2O_3 content, which could affect biogas production efficiency. The inert substances SiO_2 and Al_2O_3 might reduce nutrient availability for microorganisms engaged in anaerobic digestion. Alternatively, site B has low levels of SiO_2 and Al_2O_3, making it a desirable location for biogas production. In site B, CaO could act as a buffer and maintain

Location	Dzwerani	Tshino	Maila	Gogogo
Al_2O_3 (%)	2.2	<0.1	<0.1	<0.1
CaO (%)	2.9	2.95	7.49	3.19
Fe_2O_3 (%)	5.52	1.94	1.52	1.96
K_2O (%)	4.74	1.92	1.93	1.39
MgO (%)	1.44	0.96	1.12	1.079
MnO (%)	0.105	0.082	0.094	0.17
Na_2O (%)	<0.1	<0.1	<0.1	1.502
P_2O_5 (%)	0.958	1.67	3.86	1.21
SiO_2 (%)	41.6	28.3	16.34	24.93
TiO_2 (%)	0.427	0.247	0.141	0.143
pH	6.95	6.79	6.99	7.01
C (J/g°C)	2.97	2.79	2.21	2.49

Table 1.
Chemical compositions and specific heat capacity of cow dung.

digester pH because of its high content. It is possible to produce biogas from Site C because it contains a high level of CaO and P_2O_5. CaO could act as a buffer, and P_2O_5 could provide important nutrients for microorganisms' growth in anaerobic digestion. As MgO is well known for influencing microbial activity in the anaerobic digestion process, Site D has a high content of MgO, which might be advantageous for biogas production. The chemical composition of organic materials at different sites can significantly affect biogas production efficiency. Based on the results presented, it is important to understand the chemical composition of organic matter at different sites to determine whether biogas production is feasible. Data from the table can be used to analyze the relationship between cow dung's specific heat and the oxides present at each site. In the last column, the specific heat values are expressed in (J/g°C) and represent the specific heat of cow dung. Specifically, specific heat measures how much heat energy is required to raise a substance's temperature by a certain amount. Other factors affecting cow dung's specific heat are the organic matter's composition and the oxides it contains. Although the table does not directly provide specific heat values for individual oxides, we can still make some observations and discuss the potential effects of oxide composition on specific heat cow dung.

Dzwerani:

Dzwerani has relatively high proportions of Al_2O_3, Fe_2O_3, and SiO_2 compared to other sites. There is a possibility that these oxides contribute to the cow dung's overall specific heat. Dzwerani dung likely has a higher heat capacity because of its higher specific heat value (2.97 J/g°C); in other words, it requires more energy to warm up than other dung samples.

Tshino:

Compared to other sites, Tshino has a lower oxide content overall. Compared to Dzwerani cow dung, Tshino cow dung has a lower specific heat value (2.79 J/g°C).

Maila:

Maila has lower oxide content than Dzwerani. The cow dung from Maila appears to have a lower heat capacity based on its specific heat value of 2.21 J/g°C.

Gogogo:

There are slightly higher proportions of oxides such as CaO and MgO at the Gogogo site than at Maila. The specific heat of the cow dung from Gogogo (2.49 J/g°C) is slightly higher than the heat capacity of the cow dung from Maila. These observations can be made considering the given data and assumptions. A detailed analysis and measurement of the specific heat of each oxide present in cow dung would be required to establish a more accurate relationship between the specific heat of cow dung and its oxides. Other factors, such as moisture content, density, and impurities, may also impact cow dung-specific heat. Therefore, further experimental data and analysis are needed to determine the relationship between specific heat and oxide composition.

4. Conclusions

Based on the results of this study, it can be concluded that cow dung is characterized by its composition. Higher proportions of Al_2O_3, Fe_2O_3, and SiO_2 increase specific heat capacity. Specific heat capacity can be used to determine how much heat is required to raise a slurry's temperature to a desired operating temperature. Since anaerobic digestion is determined by cow dung and water heat and capacity, biogas production can be optimized using information about cow dung's chemical composition and specific heat capacity. Through experimentation, it was found that cow dung's chemical composition influences biogas production efficiency. Certain elements and oxides stop microorganisms from digesting cow dung.

This study could help increase agricultural production, as biogas can be used for more efficient irrigation and fertilization. This would lead to more efficient resource management and increased sustainability in rural communities.

It can be concluded that cow dung'chemical composition affects the specific heat capacity at different sites. Based on the specific heat values of cow dung at various sites, it appears to have a varying heat capacity, possibly due to the oxide composition. To assess the feasibility of biogas production, it is crucial to understand the chemical composition of organic matter at different locations. Further research and analysis are necessary for a more comprehensive understanding of the relationship between specific heat capacity and oxide composition in cow dung. This includes measurements of specific heat for individual oxides.

Acknowledgements

I would like to thank all participants who collected samples for investigation.

Conflict of interest

No conflict of interest to declare.

Author details

Vhutshilo Nekhubvi
Faculty of Science, Engineering and Agriculture, University of Venda, South Africa

*Address all correspondence to: vhutshilo.nekhubvi@univen.ac.za

References

[1] Mulu GM, Belay S, Getachew E, Tilahum WS. The environmental benefits of domestic biogas technology in rural Ethiopia. Biomass and Bioenergy. 2016;**90**:131-138. DOI: 10.1016/j.biombioe.2016.04.002

[2] Laichena JK, Wafula JC. Biogas technology for rural households in Kenya. OPEC Energy Review. 1997; **21**(3):223-244. DOI: 10.1111/1468-0076.00043

[3] Tewelde GB, Rahwa GT, Grmanesh AD, Lemlem SM. Biogas plant distribution for rural household sustainable energy supply in Africa. Energy Policy Research. 2017; **41**(1):10-20. DOI: 10.30585/easyr.2021.4.1

[4] Bhat PN, Chanakya HN. Biogas plant dissemination: Success story of Sirsi, India. Energy for Sustainable Development. 2001;**5**(1):39-46. DOI: 10.1016/S0973-0826(08)60207-5

[5] Msibi SS, Gerrit K. Potential for domestic biogas as household energy. Journal of Energy Southern Africa. 2017; **28**(2):1-13. DOI: 10.17159/2413-3051/2017/v28i2a1567

[6] Baba SUIA, Nasir I. Anaerobic digestion of cow dung for biogas production. ARPN Journal of Engineering and Applied Science. 2012; **7**(2):1819-6608. DOI: 10.9790/3021-07213854

[7] Sruthy B, Anisha GK, Gibi MM, Sruthi GR. An experimental investigation on strength of concrete made with cow dung ash and glass fiber. International Journal of Engineering and Research Technology. 2017;**6**(3): 492-495. DOI: 10.17577/IJERTV6IS030635

[8] Raja MKMM, Manne R, Devarajan A. Benefits of cow dung – A human ignored gift. Journal of Natural Remedies. 2021; **21**(3):189-202 DOI: 10.22271/j.0976-1023.2021.v20.i2a.17055

[9] Szymajda A, Łaska G, Joka M. Assessment of cow dung pellets as a renewable solid fuel in direct combustion technologies. Energies. 2021; **14**(4):1192. DOI: 10.3390/en14123322

[10] Ogur E, Irungu P. Design of a biogas generator. International Journal of Engineering Research and Applications. 2013;**3**(6):630-635. DOI: 10.13140/RG.2.1.2027.9125

[11] Alfa IM, Dahunsi SO, Iorhemen OT, Okafor CC, Ajayi SA. Comparative evaluation of biogas production from Poultry droppings, Cow dung and Lemon grass. Bioresource Technology. 2014;**157**:270-277. DOI: 10.1016/j.biortech.2014.01.066

[12] Wang H, Aguirre-Villegas HA, Larson RA, Alkan-Ozkaynak A. Physical properties of dairy manure pre- and post-anaerobic digestion. Applied Sciences. 2019;**9**(13):2-10. DOI: 10.3390/app9132703

[13] Nsair A, Cinar SO, Alassali A, Qdais HA, Kuchta K. Operational parameters of biogas plants: A review and evaluation study. Energies. 2020;**13**(15):1-27. DOI: 10.3390/en13153761

[14] Jiang J, He S, Kang X, et al. Effect of organic loading rate and temperature on the anaerobic digestion of municipal solid waste: Process performance and energy recovery. Frontiers in Energy Research. 2020;**8**. DOI: 10.3389/fenrg.2020.613848

[15] Rennuit C, Sommer SG. Decision support for the construction of farm-scale biogas digesters in developing countries with cold seasons. Energies. 2013;**6**:5314-5332. DOI: 10.3390/en6105314

[16] Obot MS, Li C, Fang T, Chen J. Measurement of thermal properties of white radish (*Raphanus raphanistrum*) using easily constructed probes. PLoS One. 2017;**12**(2):e0171016. DOI: 10.1371/journal.pone.0171016

[17] Sailer G, Silberhorn M, Eichermüller J, Poetsch J, Pelz S, Oechsner H, et al. Influence of digester temperature on methane yield of organic fraction of municipal solid waste (OFMSW). Applied Sciences. 2021;**11**:2820. DOI: 10.3390/app11062820

[18] Gebremedhin KG, Wu B, Gooch C, Wright P, Inglis S. Heat transfer model for plug-flow anaerobic digesters. American Society of Agricultural Engineers. 2005;**48**(2):777-785. DOI: 10.13031/2013.18053

[19] Nayyeri MA, Kianmehr MH, Arabhosseini A, Hassan-Beygi SR. Thermal properties of dairy cattle manure. International Agrophysics. 2009;**23**:359-366. DOI: 10.1515/AGROPH.2009.023

[20] Yerima I, Ngulde YM. Sources and uses of fuel for fish smoking around the lake Chad in Borno State, Nigeria. Journal of Environment and Ecology. 2015;**6**(2):78-93. DOI: 10.5296/jee.v6i2.8707

[21] Pvt ME. Optimization of Biogas Plants using Digester Heating Technologies. Lalitpur, Nepal: MinErgy Pvt. Ltd.; 2014

[22] Das A, Mondal C. Comparative kinetic study of anaerobic treatment of thermally pretreated source-sorted organic market refuse. Journal of Engineering. 2015;**2015**. Article ID 684749. DOI: 10.1155/2015/684749

[23] ASAE Standards. S385.2. Moisture Measurement – Forages. St. Joseph, MI, USA; 2002

[24] Ma B, Zhou X, Liu J, You Z, Wei K, Huang X. Determination of specific heat capacity on composite shape-stabilized phase change materials and asphalt mixtures by heat exchange system. Materials. 2016;**5**(9):2-15. DOI: 10.3390/ma9090685

[25] Mohsenin NN. Thermal Properties of Foods and Agricultural. New York, USA: Gordon-Breach Press; 1980

Chapter 4

Mathematical Modeling and Applied Calculation of Bioconveyer and Anaerobic Biofiltration

Vadym Poliakov

Abstract

Deep wastewater treatment is carried out by a bioconveyer technology and a direct-flow system of multistage biological treatment (DSMBT). To reduce excess biomass and energy costs, the first bioreactors (sections) of the DSMBT provide for partial removal of soluble organics in the absence of oxygen, the rest are intended for aerobic water treatment. Development of a method for engineering calculation of anaerobic biofiltration as applied to the first sections of DSMBT is the main aim of the study. Anaerobic substrate utilization is modeled at two levels using two stage biokinetics. The behavior of the substrate and its derivatives within a representative biofilm are analyzed, taking into account surface and molecular diffusion, limitation of the rates of the substrate and acid biodegradation, coexistence of two communities of microorganisms. The behavior of the original and newly formed organic substrates in the volume of a representative section is studied by analytical methods. A theoretical base and an engineering method for calculating anaerobic biofiltration are developed and illustrated, which can serve as the basis for applied optimization of the parameters of bioconveyer plants. It is justified to use the derived calculation dependencies for similar complex biological treatment plants with any filtering material.

Keywords: anaerobic biofiltration, bioconveyer, calculation method, biofilm, output concentration, substrate, utilization

1. Introduction

Deep treatment of wastewater containing a large amount of various organic contaminants has to be carried out in stages on special purification plants of complex design. Naturally, the technological process should provide at each stage a significant reduction in water pollution as a whole, but at the same time it can be aimed at the predominant removal of the dominant type of contamination. In such widespread situations, only rational technologies and structural decisions can achieve the desired result [1–8]. The bioconveyer technology and the direct-flow system of multi-stage biological treatment (DSMBT) are their successful examples. Their successful examples are the bioconveyer technology and the direct-flow system of multi-stage biological treatment (DSMBT). The specified technology and complex plant were proposed, and then tested on an industrial scale by professor P.I. Gvozdyak (Institute of Colloid Chemistry and Water Chemistry of the National Academy of Sciences of Ukraine) [9, 10].

The original shape and high density of structural elements contribute to the high efficiency of the new generation technology. Thus, in fact the solid phase consists of a set of vertically and densely arranged (upper end is fixed) thin synthetic profiled fiber-nozzles (viy). Their surface has a complex configuration, which contributes to the strong attachment of microorganisms to them together with their metabolites. Due to the multiplicity and vertical orientation of the fibers, the space, that is filled by them in the sequence of bioreactors (sections) of DSMBT, has a high hydraulic resistance and anisotropy. Therefore, it is justified to interpret it as a specific (pseudo) porous medium. Simultaneously, the water flow between fibers is conditionally horizontal. This fact gives the right to use one-dimensional mathematical models of hydrodynamics in anisotropic porous medium, mass transfer and take into account changes in the permeability of the medium only in this direction.

Currently, there are acute problems of removing xenobiotics, disposal of waste from biological water treatment. It is possible to significantly reduce the amount of excess biomass, as well as to minimize energy costs, due to the decomposition of organic contaminants in the absence of oxygen. An appropriate technological process is implemented (in fact, low-molecular easily degradable compounds are processed) as a result of the successive and parallel course of a series of (bio)chemical reactions [11–13]. Detailed biokinetic model includes nine equations [14]. However, when developing engineering methods for calculating anaerobic biofiltration, it is sufficient to single out and formalize those processes and effects that determine the operation of the corresponding bioreactor-filter and, most importantly, the final result. The justification for the simplified consideration of the specified complex technological process as a two-stage process, which is just implemented below, is presented in the works [15, 16]. Taking into account the specifics of water purification by anaerobic microbiocenosis in a porous medium, its productive potential deserves special attention, in fact, the growth rate of anaerobes population under the most favorable conditions [17–19]. Actually, the vital activity of any microorganisms, including anaerobes, is significantly limited under the influence of many negative factors of a physical, chemical, biological nature [20–24]. Low content of the initial and especially intermediate (long-chain fatty acids) substrates may be the main limiting factor [25–27]. The adequacy of the initial model to the real conditions of biofiltration, its provision with reliable information are of great importance in mathematical modeling of the utilization of easily mineralizable organics in anaerobic bioreactor-filters. The source of such information are usually the articles that contain original data or data borrowed from experimental and theoretical studies, for example [28–30].

The principles of organizing the most complex technological process at DSMBT (deep biological treatment of highly toxic waters) are illustrated in **Figure 1**. It is essential that the bioconveyer technology is able to purify both relatively clean and extremely polluted water, while being waste-free. However, such a result can only be achieved with scientifically based debugging of DSMBT, namely, primarily due to the selection of populations of microorganisms that are most suitable for solving local applied problems. The presented three-stage layout of the technology of biological water treatment guarantees the production of high-quality drinking water with valuable biological additives. Developed on this base, the calculation methods make it possible to reliably assess the consequences of the action of DSMBT at the stages of microbiological treatment. The calculation method in relation to the second (aerobic)

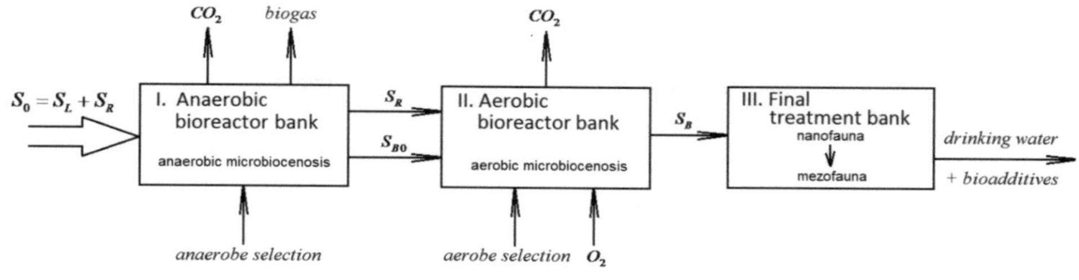

Figure 1.
Schematic diagram to calculate biological treatment at DSMBT: S_O – Initial concentration of organic pollution; S_L, S_R – Concentrations of its low-molecular and high-molecular components; S_{BO} – Concentration of primary bacterial pollution;S_B - concentration of total bacterial pollution.

stage in the first version is set out in the articles [31, 32]. The method applied to the anaerobic stage is presented and discussed below. At the same time, these methods can serve as the basis for the applied optimization of the technological and design parameters of the first bioreactor banks (anaerobic I and aerobic II), and in the future, the entire purification plant. However, for full-scale calculations of DSMBT, it is also necessary to model the life activity of coexisting organisms that continue the trophic food chain - from protozoa to filter feeders and then predators. At present, modeling the removal of bacterial pollution, and at the same time, the waste products of the nanofauna, its mobilized individuals is problematic primarily due to the lack of suitable experimental information.

The soluble organic component of pollution is finally eliminated and the content of high-molecular compounds drops to almost 0 in the bank II. In order to avoid organic shock load on it (can cause an emergency), an anaerobic reactor bank I is equipped (**Figure 1**). A comparative analysis of water treatment by aerobic and anaerobic microbiocenoses was carried out in the work [33]. At the same time, conditions are created in the bank III (usually includes two bioreactors) that favor the vital activity of organisms with more complex physiology (protozoa, filter feeders, predators). The first section (bioreactor) of the bank is inhabited by representatives of the nanofauna, while the second section is inhabited by the mesofauna. Of particular note is the ease of operation and flexibility in the management of DSMBT, its compactness (if we take into account the scale of the plant's activity). Indeed, the service of the plant is comparatively easy. The plant is able to independently adapt to extreme conditions. If necessary, it is easy to forcibly adjust the composition of the microbiocenosis using specially selected microorganisms, to supply nutrients in the required amount.

Substrate biodegradation in different sections of the same bank is described by models of the same type, similar coefficients of which may differ significantly. Since the diffusion mechanism of mass transfer on the scale of each section can be neglected [34], the relationship between neighboring sections and banks (by analogy with a layered medium of a rapid filter, multi-stage filters) is almost one-sided. As a result, the values of pollution concentrations at the outlet of each section, excluding the last section, can be considered as initial values when modeling biofiltration. Hereinafter, the term biofiltration will be systematically used, although it does not quite correspond to the real conditions in the specific medium under consideration.

Thus, the algorithm for calculating the action of DSMBT is actually reduced to the sequential determination of the output concentrations of substrates for the sections of the first, second and, as a result, the third bioreactor bank. Therefore, in order to develop a method for calculating DSMBT, it is first necessary to analyze, using analytical methods, the utilization of the organic substrate in representative anaerobic and aerobic banks (sections) under nonuniform boundary conditions. Similar studies in relation to aerobic conditions were carried out earlier in the above-mentioned works of the author.

It should be emphasized that the anaerobic biofiltration, organized within one or two sections of the first bank of DSMBT, as well as the previously studied the aerobic biofiltration, are modeled in a similar way, namely, at two levels. A key role in the anaerobic utilization of dissolved organic matter is also played by biofilms which differ significantly in size and properties (composition, adaptability, activity, strength, etc.) from aerobic analogues. And above all, due to the relatively slow growth of anaerobic microorganisms, the current (l_f) and limiting (l_{fm}) thicknesses of the biofilm formed by them, as a rule, are clearly less than those of aerobic biofilms. Therefore, when choosing the radius of the fiber (R_t) as a linear scale, the following relation $\bar{l}_{fm} = l_{fm}/R_t < < 1$ also holds.

It is appropriate to note that this development can be useful in the case of an in-depth study of the action of biofilms with poor aeration of polluted waters and, as a result, the localization of oxygen in their outer part (aerobic zone). However, the internal part of the biofilm (anaerobic zone) can also make a noticeable contribution to the processing of the organic substrate. Of course, ignoring such a contribution is uncritical in view of its usually relative smallness. However, for the operation of the first bank of the DSMBT, the vital activity of anaerobes is of decisive importance for the operation of the first bank of the DSMBT. Therefore, the attention is focused below precisely on the steady-state (due to the stable long-term operation of this plant) anaerobic biofiltration [35, 36].

2. Theoretical analysis

2.1 Statement and solution of mathematical problem

Thus, the behavior of the substrate and its derivatives within an arbitrary flat biofilm with a thickness l_f (in view of $l_f < < R_t$) is analyzed by analytical methods at the first stage of studies of stationary anaerobic biofiltration in a (pseudo) porous medium. It is here that it transforms in several stages, so that the end products are volatile acids (secondary substrate), carbon dioxide and combustible biogas, mainly methane. In fact, two stages dominate, at which these products are formed. The internal mathematical problem in this case is formulated with respect to the corresponding mass concentrations $s_i \, (i = 1, 2, 3, 4)$ assuming only diffusion (surface + molecular diffusion) mass transfer, significant limitation of the biodegradation rates of the substrate and acids, coexistence within a single biofilm of two communities of microorganisms (acid-producing and methane-producing). The problem includes, first of all, the following system of equations

$$D_{e1}\frac{d^2 s_1}{dx^2} = \frac{1}{Y_1}\frac{\mu_{m1}\rho_{B1}s_1}{K_{s1} + s_1}, \quad (1)$$

$$D_{e2}\frac{d^2 s_2}{dx^2} = \frac{1}{Y_2}\frac{\mu_{m2}\rho_{B2}s_2}{K_{s2} + s_2} - \frac{\mu_{m1}\rho_{B1}Y_{s_2/B_1}s_1}{K_{s1} + s_1}, \quad (2)$$

$$D_{e3}\frac{d^2s_3}{dx^2}=\frac{\mu_{m1}\rho_{B1}Y_{s_3/B_1}s_1}{K_{s1}+s_1}+\frac{\mu_{m2}\rho_{B2}Y_{s_3/B_2}s_2}{K_{s2}+s_2}, \tag{3}$$

$$D_{e4}\frac{d^2s_4}{dx^2}=\frac{\mu_{m2}\rho_{B2}Y_{s_4/B_2}s_2}{K_{s2}+s_2}. \tag{4}$$

Here, D_{ei} is the effective diffusion coefficient of the i-th substance; $Y_{1,2}$ are the effective economic coefficients that characterize the decomposition of the initial substrate and volatile acids by the corresponding groups of microorganisms in general. Also.

$$Y_1=\frac{1}{Y_{s_1/B_1}+Y_{s_2/B_1}+Y_{s_3/B_1}}, Y_2=\frac{1}{Y_{s_{21}/B_2}+Y_{s_3/B_2}+Y_{s_4/B_2}}; \tag{5}$$

Y_{s_i/B_j} are the conversion coefficients equal to the mass of the formed or consumed substance, which falls on the biomass unit of the j-th variety; μ_{mj}, ρ_{Bj} are the specific growth rate and density of the j-th biomass. This system is supplemented with standard boundary conditions for biofilms (on the surfaces separating the biofilm from the liquid film and fiber-nozzle).

$$x=0, \frac{ds_i}{dx}=0; x=l_f, D_{ei}\frac{ds_i}{dx}=k_{Li}(S_i-s_i); \tag{6}$$

where k_{Li} is the transfer coefficient of the i-th substance through the liquid film; S_i is the concentration of the i-th substance in the liquid medium outside both films.

Dimensionless variables and parameters are introduced using as scales S_{10} (concentration of the primary substrate at the inlet to the section under consideration), R_t (characteristic microsize), D_{e1} as follows:

$\overline{S}_i, \overline{s}_i=\frac{S_i, s_i}{S_{10}}, \overline{x}=\frac{x}{R_t}, \overline{\lambda}_j=\frac{\mu_{mj}\rho_{Bj}\lambda_j}{Y_jD_{e1}S_{10}}, \overline{D}_e=\frac{D_{e2}}{D_{e1}}, \overline{K}_{si}=\frac{K_{si}}{S_{10}}, \overline{k}_{Li}=\frac{R_tk_{Li}}{D_{ei}}, \overline{l}_f=\frac{l_f}{R_t}$. If the theoretical analysis of anaerobic biofiltration is performed solely for the purpose of monitoring the quality of biological water treatment, and the combustible gas released along the way is of no practical interest, then it is enough to restrict ourselves to a truncated system of the equations for $\overline{s}_1, \overline{s}_2$, namely,

$$\frac{d^2\overline{s}_1}{d\overline{x}^2}=\frac{\overline{\lambda}_1\overline{s}_1}{\overline{K}_{s1}+\overline{s}_1}, \tag{7}$$

$$\frac{d^2\overline{s}_2}{d\overline{x}^2}=\frac{\overline{\lambda}_2\overline{s}_2}{\overline{K}_{s2}+\overline{s}_2}-\frac{Y_1Y_{s2/B_1}}{\overline{D}}\frac{\overline{\lambda}_1\overline{s}_1}{\overline{K}_{s1}+\overline{s}_1} \tag{8}$$

and the boundary conditions.

$$\overline{x}=0, \frac{d\overline{s}_{1,2}}{s\overline{x}}=0; \tag{9}$$

$$\overline{x}=\overline{l}_f, \frac{d\overline{s}_{1,2}}{d\overline{x}}=\overline{k}_{L1,L2}\left(\overline{S}_{1,2}-\overline{s}_{1,2}\right). \tag{10}$$

The solution of problem (7)–(10) can be significantly simplified due to the usually large initial content of the substrate in wastewater. In such situations, which are just characteristic of bioconveyer technologies, it is reasonable to assume that the primary substrate decomposes at a maximum rate. It should be noted that this assumption does not apply to volatile acids, the decomposition of which is not only limited due to their low content. At the same time, an inhibitory effect is also possible. Therefore, the indicated maximum rate will be $\overline{\lambda}_1$ and Eq. (8) is reduced to the form

$$\frac{d^2\overline{s}_2}{d\overline{x}^2} = \frac{\overline{\lambda}_2\overline{s}_2}{\overline{K}_{s2} + \overline{s}_2} - \tilde{\lambda}_1, \tag{11}$$

where $\tilde{\lambda}_1 = Y_1 Y_{s_2/B_1}\overline{\lambda}_1/\overline{D}$. In that case the distribution of the primary substrate across the biofilm is represented by the following expression

$$\overline{s}_1(\overline{x}) = \overline{S}_1 - \frac{\tilde{\lambda}_1\overline{l}_f}{\overline{k}_{L1}} + \frac{\tilde{\lambda}_1}{2}\left(\overline{x}^2 - \overline{l}_f^2\right). \tag{12}$$

Problem (9)–(11) is approximately solved by averaging the right side of Eq. (11). Previously, this technique was used in the absence of an internal source of degradable substrate $(\tilde{\lambda}_1 = 0)$ and then substantiated on the test examples [31]. Therefore, following the previous procedure and keeping the notation, we derive an equation for the average value u_{av} on the interval $[0, X]$

$$u_{av}(\mathrm{X}) = \frac{1}{\mathrm{X}}\int_0^{\mathrm{X}} \frac{\overline{s}_2(\overline{x})d\overline{x}}{\overline{K}_{s2} + \overline{s}_2(\overline{x})}, \tag{13}$$

which looks like

$$u_{av} + \frac{2\overline{K}_{s2}}{\left(\overline{\lambda}_s u_{av} - \tilde{\lambda}_1\right)\cdot \Psi_f(u_{av})\overline{x}} arctg \frac{\overline{x}}{\Psi_f(u_{av})} = 1. \tag{14}$$

Here $\Psi_f^2 = \frac{2\left(\overline{K}_{s2}+\overline{S}_2\right)}{\overline{\lambda}_2 u_{av} - \mathrm{X}_1} - \frac{2\overline{D}\overline{l}_f}{\overline{k}_{L2}} - \overline{l}_f^2$. Now the function $arctg\frac{\overline{x}}{\Psi_f}$ is expanded into a series in terms of the argument and only its first term is preserved. The result is a quadratic equation for u_{av}

$$\overline{\lambda}_2\phi_f\left(\overline{l}_f\right)u_{av}^2 - \left[\overline{\lambda}_2\phi_f\left(\overline{l}_f\right) + \overline{\lambda}_2\overline{K}_{s2} + \overline{\lambda}_2\overline{S}_2 + \tilde{\lambda}_1\phi_f\left(\overline{l}_f\right)\right]u_{a2} + \tilde{\lambda}_1\phi_f\left(\overline{l}_f\right) + \overline{\lambda}_2\overline{S}_2 = 0, \tag{15}$$

where $\phi_f\left(\overline{l}_f\right) = \overline{\lambda}_2\left(\overline{k}_{L2}\overline{l}_f^2 + 2\overline{D}\overline{l}_f\right)/2\overline{k}_{L2}$. Physical meaning has only one root, namely,

$$u_{av}\left(\overline{l}_f, \overline{S}_2\right) = \frac{1}{2}\left\{\frac{\overline{K}_{s2} + \overline{S}_2}{\phi_f\left(\overline{l}_f\right)} + \frac{\tilde{\lambda}_1}{\overline{\lambda}_2} + 1 - \sqrt{\left(\frac{\overline{K}_{s2} + \overline{S}_2}{\phi_f\left(\overline{l}_f\right)} + \frac{\tilde{\lambda}_1}{\overline{\lambda}_2} + 1\right)^2 - \frac{4\overline{S}_2}{\phi_f\left(\overline{l}_f\right)} - \frac{4\tilde{\lambda}_1}{\overline{\lambda}_2}}\right\}. \tag{16}$$

After the simplification of Eq. (11) considering Eq. (13) and then its double integration, the expression for the concentration $\overline{s}_2$ is derived

$$\overline{s}_2\left(\overline{x}; \overline{l}_f, \overline{S}_2\right) = \overline{S}_2 - \left[\overline{\lambda}_2 u_{av}\left(\overline{l}_f, \overline{S}_2\right) - \tilde{\lambda}_1\right]\left(\frac{\overline{l}_f}{\overline{k}_{L2}} + \frac{\overline{l}_f^2}{2} - \frac{\overline{x}^2}{2}\right). \tag{17}$$

Thus, the relative value $\overline{s}_2$ on the outer surface of the biofilm will be

$$\bar{s}_{2f}\left(\bar{l}_f, \bar{S}_2\right) = \bar{S}_2 - \left[\tilde{\lambda}_2 u_{av}\left(\bar{l}_f, \bar{S}_2\right) - \tilde{\lambda}_1\right] \frac{\bar{l}_f}{\bar{k}_{L2}}. \tag{18}$$

Using the representations for $\bar{s}_1$ Eq. (12) and $\bar{s}_2$ Eq. (17) with the help of double integrating Eqs. (3) and (4) under the appropriate conditions Eq. (6) it is also easy to find the concentrations $\bar{s}_3, \bar{s}_4$ of the main end products of decomposition (CO_2, CH_4).

To assess the actual active capacity of the biofilm in relation to both components of organic pollution allows the calculation of their relative rates$\bar{i}_{f1}$, $\bar{i}_{f2}$ across the boundary between both films

$$\bar{i}_{f1}\left(\bar{l}_f\right) = \tilde{\lambda}_1 \bar{l}_f, \tag{19}$$

$$\bar{i}_{f2}\left(\bar{l}_f, \bar{S}_2\right) = \bar{l}_f\left[\tilde{\lambda}_2 u_{av}\left(\bar{l}_f, \bar{S}_2\right) - \tilde{\lambda}_1\right]. \tag{20}$$

At the second stage of theoretical studies of steady-state anaerobic biofiltration, attention is focused on the behavior of the initial and newly formed (secondary) organic substrates on the scale of the section selected for consideration. The basis of the corresponding mathematical model (the biofiltration compartment in the general model) is the system of equations for the convective transfer of both substrates within the given section. Neglecting the diffusion (dispersion) mechanism, the system of macrotransfer equations can be written in the dimensional form

$$V\frac{dS_1}{dy} + I_{B1}\left(l_f\right) = 0, \tag{21}$$

$$V\frac{dS_2}{dy} + I_{B2}\left(l_f, S_2\right) = 0, \tag{22}$$

where y is the coordinate in the direction of water movement. Here, the functions of utilization of primary $I_{B1}\left(l_f\right)$ and secondary $I_{B1}\left(l_f\right)$ substrates at the specific surface area of the biological phase Ω_B are presented, for example, in the following form.

$$I_j = \Omega_B D_{ej} \frac{ds_j}{dx}\Big|_{x=l_j}; \; j = 1, 2 \tag{23}$$

In the analyzed case of bioconveyer technology and the anaerobic section with biomass at the fibers of a radius R_t, the area Ω_B is expressed in terms of the fraction of space free from them n_0 (analogous to porosity for porous media) as follows

$$\Omega_B = 2(1 - n_0)/R_t. \tag{24}$$

A similar expression as applied to a medium of grains with a radius R_g will be

$$\Omega_B = 3(1 - n_0)/R_g. \tag{25}$$

Equations (21) and (22) are supplemented with the boundary conditions.

$$y = 0, S_j = S_{j0}. \tag{26}$$

To establish a relationship between the relative thickness $\bar{l}_f$ and concentration of the total biomass B as the sum of B_1 (acid-producing bacteria and their metabolites) and B_2 (methane-producing bacteria and their metabolites), the generalized balance equation between its increase and decrease is used

$$\Upsilon_{B_1/s_1} I_1 + \Upsilon_{B_2/s_2} I_2 = k_d(B_1 + B_2) = k_d B. \tag{27}$$

Equation (27) are transformed after a series of transformations to this equality in the dimensionless form

$$u_{av}\left(\bar{l}_f, \bar{S}_2\right) = \psi_B = \frac{1}{\bar{\lambda}_2}\left[\bar{k}_{d2}(\bar{\rho}_B + 1) - \frac{\bar{\lambda}_1}{\bar{D}}\left(\frac{\Upsilon_{B_1/s_1}}{\Upsilon_{B_2/s_2}} - \Upsilon_1 \Upsilon_{s_2/B_1}\right)\right], \tag{28}$$

where $\bar{k}_{d2} = k_d \rho_{B2} \Upsilon_{s_2/B_2} R_t^2 / (D_{e2} S_{10})$, $\bar{\rho}_B = \rho_{B1}/\rho_{B2}$. An expression is derived from Eq. (28) for the concentration $\bar{S}_2$ as a function of $\bar{l}_f$ taking into account Eq. (16)

$$\bar{S}_2\left(\bar{l}_f\right) = \frac{\bar{\lambda}_2 \psi_B - \tilde{\lambda}_1}{\bar{\lambda}_2} \phi_f\left(\bar{l}_f\right) + \frac{\psi_B \overline{K}_{s2}}{1 - \psi_B}. \tag{29}$$

Of fundamental importance for modeling biofiltration is the assumption of a weak dependence of the rate of biomass loss on the concentration of dissolved organic matter, which allows us to assume $k_d = k_d\left(l_f\right)$. From a formal point of view, it is much easier when describing the utilization of substrates in the bioreactor medium to operate instead of $I_j$using the equivalent expressions from the right side of Eq. (27). Then system Eqs. (21) and (22) takes the following form

$$\frac{d\bar{S}_1}{d\bar{y}} + \chi_t \bar{k}_{d1} \bar{l}_f = 0, \tag{30}$$

$$\frac{d\bar{S}_2}{d\bar{y}} + \chi_t \overline{D} \bar{k}_{d2} \bar{l}_f = 0, \tag{31}$$

where $\chi_t = 2(1 - n_0)\frac{D_{e1} L}{V R_t^2}$ (for fiber-nozzles), $\bar{k}_{d1} = \frac{k_d \rho_{B1} R_t^2}{\Upsilon_{s_1/B_1} D_{e1} S_{10}}$.

The subsequent analysis by analytical methods of the general stabilized situation in the reactor medium and the choice of an appropriate calculation scheme are determined by the degree of saturation with biomass of the space between the fibers. Since the behavior of the secondary substrate is much more difficult to formalize, therefore, the attention will be given specially to it.

The determination of the relative thickness of biofilms in the inlet section of the medium l_{f0} is of key importance for concretizing the situation. Here it is possible to express $\bar{l}_f$ through $\bar{S}_{20}$ and finally derive the formula

$$\bar{l}_{f0} = \sqrt{\frac{\overline{D}^2}{\bar{k}_L^2} + \frac{2}{\bar{\lambda}_2 \psi_B - \tilde{\lambda}_1}\left(\bar{S}_{20} - \frac{\psi_B \overline{K}_{s2}}{1 - \psi_B}\right)} - \frac{\overline{D}}{\bar{k}_L}. \tag{32}$$

Then $\bar{l}_{f0}$ is correlated with the limit value $\bar{l}_{fm}$. The first two more realistic situations turn out if $l_{f0} > l_{fm}$. Equations (30) and (31) can be easily integrated in this case taking into consideration the boundary conditions.

$$\bar{y} = 0, \bar{S}_1 = 1; \bar{S}_2 = \bar{S}_{20} \tag{33}$$

As a result, the following linear representations are obtained for the desired concentrations

$$\overline{S}_1(\overline{y}) = 1 - \chi_t \overline{k}_{d1} \overline{l}_{fm} \overline{y}, \tag{34}$$

$$\overline{S}_2(\overline{y}) = \overline{S}_{20} - \chi_t \overline{D}\overline{k}_{d2} \overline{l}_{fm} \overline{y}. \tag{35}$$

The next computational step is to calculate the value $\overline{S}_{2m}$ (using Eq. (29) for given values $\overline{l}_{fm}, \overline{S}_{20}$) related to

$$\overline{S}_{2m} = \left(\psi_B - \frac{\tilde{\lambda}_1}{\overline{\lambda}_2}\right)\phi_f\left(\overline{l}_{fm}\right) + \frac{\psi_B \overline{K}_{s2}}{1-\psi_B}. \tag{36}$$

Then the coordinate $\overline{y}_m$ of the section, in which $\overline{S}_2 = \overline{S}_{2m}$, is calculated by the formula following from Eq. (35),

$$\overline{y}_m = \frac{\overline{S}_{20} - \overline{S}_{2m}}{\chi_t \overline{D}\overline{k}_{d2} \overline{l}_{fm}}. \tag{37}$$

Two situations are possible depending on the ratios $\overline{y}_m \gtrless 1$. First, at $\overline{y}_m > 1$ (the entire medium contains the maximum amount of biomass) the most important for practice output concentrations $\overline{S}_{ie}$ are simply calculated from (34) and (35).

$$\overline{S}_{ie} = 1 - \chi_t \overline{k}_{d1} \overline{l}_{fm}, \overline{S}_{2e} = \overline{S}_{20} - \chi_t \overline{D}\overline{k}_{d2} \overline{l}_{fm} \tag{38}$$

Calculations of anaerobic biofiltration are much more difficult if $1 > \overline{y}_m > 0$. Then two characteristic zones are formed in the medium, where, respectively, $l_f = l_{fm}$ and $l_f < l_{fm}$. Within the first zone, as before, linear distributions Eqs. (34) and (35) are valid up to the section $y = y_m$. In the second zone $(1 \geq y \geq y_m)$, these distributions are already non-linear due to decreasing l_f along the flow. The corresponding distribution $\overline{l}_f(\overline{y})$ is found from Eq. (31), which is transformed to this form

$$\frac{d}{d\overline{l}_f}\overline{S}_2\left(\overline{l}_f\right) \cdot \frac{d\overline{l}_f}{d\overline{y}} + \chi_t \overline{D}\overline{k}_{d2} \overline{l}_f = 0 \tag{39}$$

and solved under the condition.

$$\overline{y} = \overline{y}_m, \overline{l}_f = \overline{l}_{fm}. \tag{40}$$

Based on Eq. (31), an expression for $d\overline{S}_2/d\overline{l}_f$ is derived and then a solution to problem Eqs. (39) and (40) is obtained in the form of an inverse function

$$\overline{y} - \overline{y}_m = \frac{1}{\chi_t \overline{D}\overline{k}_{d2}}\left[\frac{\overline{S}_2\left(\overline{l}_{fm}\right)}{\overline{l}_{fm}} - \frac{\overline{S}_2\left(\overline{l}_f\right)}{\overline{l}_f} + \int_{\overline{l}_f}^{\overline{l}_{fm}} \frac{\overline{S}_2(\zeta)}{\zeta^2} d\zeta\right]. \tag{41}$$

Expression Eq. (41) can be simplified after integration taking into account Eq. (29) and some transformations, so that

$$\bar{y} - \bar{y}_m = \frac{1}{\chi_t \overline{D}\bar{k}_{d2}} \left[\left(\bar{\lambda}_2 \psi_B - \bar{\lambda}_1\right)\left(\bar{l}_{fm} - \bar{l}_f\right) + \frac{\overline{D}}{\bar{k}_{L2}} \ln \frac{\bar{l}_{fm}}{\bar{l}_f} \right]. \tag{42}$$

Now, in order to determine the value $\overline{S}_2$ at any value $\bar{y}$ within the second zone, it is enough to attach Eqs. (29)–(42). Therefore, there is a parametric representation for $\overline{S}_2(\bar{y})$ and the thickness $\bar{l}_f$ is here the parameter, which decreases from $\bar{l}_{fm}$ to $\bar{l}_{fe}$. Calculations of $\overline{S}_2(\bar{y})$ can be simplified by getting rid of $\bar{l}_f$ due to the use of the dependence $\bar{l}_f\left(\overline{S}_2\right)$. This dependence has the form Eq. (32), where $\bar{l}_{f0}$ and $\overline{S}_{20}$ are formally replaced by $\bar{l}_f$ and $\overline{S}_2$. As a result, the desired distribution $\overline{S}_1(\bar{y})$ is described by the inverse function $\bar{y}\left(\overline{S}_2\right)$.

To ascertain the distribution $\overline{S}_1(\bar{y})$ in the same zone, first of all, Eq. (30) is presented as follows

$$\frac{d\overline{S}_1}{d\bar{l}_f} + \chi_t \bar{k}_{d1} \bar{l}_f \frac{d}{d\bar{l}_f} \bar{y}\left(\bar{l}_f\right) = 0. \tag{43}$$

The corresponding boundary condition will be.

$$\bar{l}_f = \bar{l}_{fm}, \overline{S}_1 = \overline{S}_{1m} = 1 - \chi_t \bar{k}_{d1} \bar{l}_{fm} \bar{y}_m. \tag{44}$$

Since according to Eq. (39)

$$\frac{d\bar{y}}{d\bar{l}_f} = -\frac{1}{\chi_t \overline{D}\bar{k}_{d2} \bar{l}_f} \frac{d\overline{S}_2}{d\bar{l}_f}, \tag{45}$$

then integration of Eq. (43) taking into account Eqs. (44) and (45) gives

$$\overline{S}_1\left(\bar{l}_f\right) = \overline{S}_{1m} + \frac{\bar{k}_{d1}}{\overline{D}\bar{k}_{d2}} \left[\overline{S}_2\left(\bar{l}_f\right) - \overline{S}_{2m}\right]. \tag{46}$$

Consequently, the relative concentration of dissolved organic matter at the outlet of the biofilter $\overline{S}_e$ will be

$$\overline{S}_e = \overline{S}_{1e} + \overline{S}_{2e} = \overline{S}_{1m} - \frac{\bar{k}_{d1}}{\overline{D}\bar{k}_{d2}} \overline{S}_{2m} + \left(1 + \frac{\bar{k}_{d1}}{\overline{D}\bar{k}_{d2}}\right) \overline{S}_{2e}, \tag{47}$$

where $\overline{S}_{2e}$ is calculated by Eq. (29) with the value $\bar{l}_{fe}$ previously found by selection from the equation

$$\left(\bar{\lambda}_2 \psi_B - \bar{\lambda}_1\right)\left(\bar{l}_{fm} - \bar{l}_{fe}\right) + \frac{\overline{D}}{\bar{k}_{L2}} \ln \frac{\bar{l}_{fm}}{\bar{l}_{fe}} = \chi_t \overline{D}\bar{k}_{d2}\left(1 - \bar{y}_m\right). \tag{48}$$

The third situation with its characteristic relation $\bar{l}_{f0} \leq \bar{l}_{fm}$ is largely similar to the second situation $\left(1 > y_m > 0\right)$. Its calculation with minimal differences duplicates the computational procedure described above. Since here there is no zone of maximum saturation with biomass at all, then $\bar{l}_f$ starting with the value $\bar{l}_{f0}$ becomes smaller according to Eq. (32) as the distance from the inlet section increases. With known $\bar{l}_{f0}$ are sequentially calculated:

biofilm thickness at the biofilter outlet from the equation

$$\left(\bar{\lambda}_2\psi_B - \tilde{\lambda}_1\right)\left(\bar{l}_{f0} - \bar{l}_{fe}\right) + \frac{\overline{D}}{\bar{k}_{L2}} \ln \frac{\bar{l}_{f0}}{\bar{l}_{fe}} = \chi_t \overline{D}\bar{k}_{d2}, \tag{49}$$

the output concentration of the secondary substrate

$$\overline{S}_{2e} = \left(\bar{\lambda}_2\psi_B - \tilde{\lambda}_1\right)\left(\frac{\bar{l}_{fe}^2}{2} + \frac{\overline{D}}{\bar{k}_{L2}}\bar{l}_{fe}\right) + \frac{\psi_B \overline{K}_{s2}}{1-\psi_B}, \tag{50}$$

the output concentration of the primary substrate

$$\overline{S}_{1e} = 1 - \frac{\bar{k}_{d1}}{\overline{D}\bar{k}_{d2}}\left(\overline{S}_{20} - \overline{S}_{2e}\right). \tag{51}$$

Obviously, when polluted water passes through the section of DSMBT under study, it is realistic to reduce the concentration of dissolved organic matter by a relative value

$$\Delta\overline{S} = 1 + \overline{S}_{20} - \overline{S}_{1e} - \overline{S}_{2e}. \tag{52}$$

2.2 Calculation of test examples and discussion of results

The approximate solutions obtained above for stationary internal (biofilm) and external (bioreactor medium) problems are illustrated by the test examples. Possible inaccuracies in the calculation of micro characteristics due to the use of adaptive averaging of the local function of the organic substrate utilization in relation to the aerobic biofilm were evaluated. It was found that they do not exceed a few percent and, as a rule, are noticeably smaller than the errors that occurred due to the experimental determination of the model coefficients. The relative flow rates of both substrates through the surface of the representative flat biofilm $\left(i_{f1}, i_{f2}\right)$ were the subject of numerous calculations. They determine the active capacity of the elements of the biological phase (biofilms of any shape) in relation to organic pollution and underlie the modeling of anaerobic biofiltration. Calculations were performed using Eqs. (19) and (20) with a continuous change in the relative thickness $\bar{l}_f$ from 0 to 0.2. Thus, the range of its real values was covered with a large margin. The initial content of volatile acids was also discretely varied from 0 to 0.5. In the adopted model, a stable presence of volatile acids is actually allowed already at the inlet to the bioreactor $(S_{20} > 0)$. In practice, such a situation is typical, along with DSMBT also for sequentially operating second and subsequent anaerobic bioreactors (bank I). The initial information included the following fixed relative values of the coefficients: $\overline{K}_{s2} = 0.25, \bar{k}_{L2} = 10, \bar{\rho}_B = \overline{D} = 1, \chi = 0.4, \bar{\lambda}_1 = 10$. Also, two characteristic values (10 and 20) were chosen for $\tilde{\lambda}_1$. Graphs of the dependence $\bar{i}_{f2}\left(\bar{l}_f\right)$ for $\tilde{\lambda}_1 = 10$ are presented in **Figure 2** and for $\tilde{\lambda}_1 = 20$ - in **Figure 3**. Here, the only graph for $\bar{i}_{f1}\left(\bar{l}_f\right)$ is given due to the constancy of $\tilde{\lambda}_1$.

When determining the values $\bar{i}_{f1}$, $\bar{i}_{f2}$, their sign is of fundamental importance, since it governs the direction of transfer of the corresponding substrate. The "+" sign means that the impurity moves inside the biofilm, and the "-" sign means - in the opposite direction. Obviously, the primary substrate is only consumed by the biofilm, and therefore the consumption rate $\bar{i}_{f1}$ is necessarily positive. The orientation of the secondary substrate is dictated by the ratio between its concentrations outside both films S_2 and at their common boundary s_{2f}. Thus, volatile acids will diffuse from the outside at $s_{2f} < S_2$ (**Figure 2** and curve 2 in **Figure 3**), and $\bar{i}_{f2}$ will already be negative at $s_f > S_2$ (curves 3-5 in **Figure 3**).

Therefore, thanks to the solution to the problem of the action of a representative anaerobic biofilm, in essence, a theoretical basis has been developed for subsequent studies using analytical methods for the operation of purification plants for biological treatment under anaerobic conditions. Also, the derived dependencies can be used to specify the model coefficients at the microlevel.

In the second series of examples, the subject of calculations were relative macrocharacteristics – biofilm thickness, which can be interpreted as a reduced biomass, concentrations of primary and secondary substrates. A larger and less significant part of the initial information for the analysis of the technological process was common to all the examples mentioned and accepted only in a dimensionless form. The indicated information included: $\overline{K}_{s2} = 0.1$, $\overline{k}_{L2} = 10$, $\overline{\rho}_B = \overline{D} = 1, \chi_t = 0.4$. The coefficients $\overline{\lambda}_i, \overline{k}_{di}$ $(i = 1, 2)$, controlling the amount of the biological phase in the bioreactor medium, varied continuously $(\overline{k}_{di})$ or discretely. In order to reduce the amount of calculations, it was assumed that $\overline{\lambda}_1$ and $\overline{\lambda}_2$, $\overline{k}_{d1}$ and $\overline{k}_{d2}$ are identical. In parallel, two options for saturation of this medium with biomass were considered. Formally, they differ in the ratio between the maximum (at the inlet to the medium)

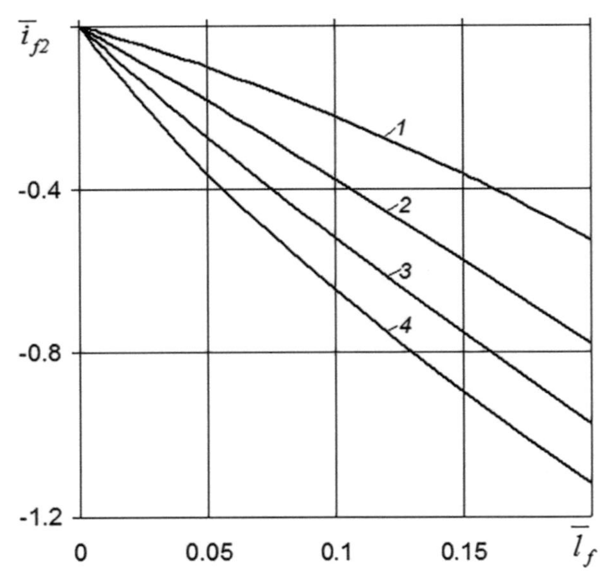

Figure 2.
Dependence $\bar{i}_{f2}(\bar{l}_f)$: $1 - \overline{S}_{20} = 0$, $2 - \overline{S}_{20} = 0.1$, $3 - \overline{S}_{20} = 0.25$, $4 - \overline{S}_{20} = 0.5$.

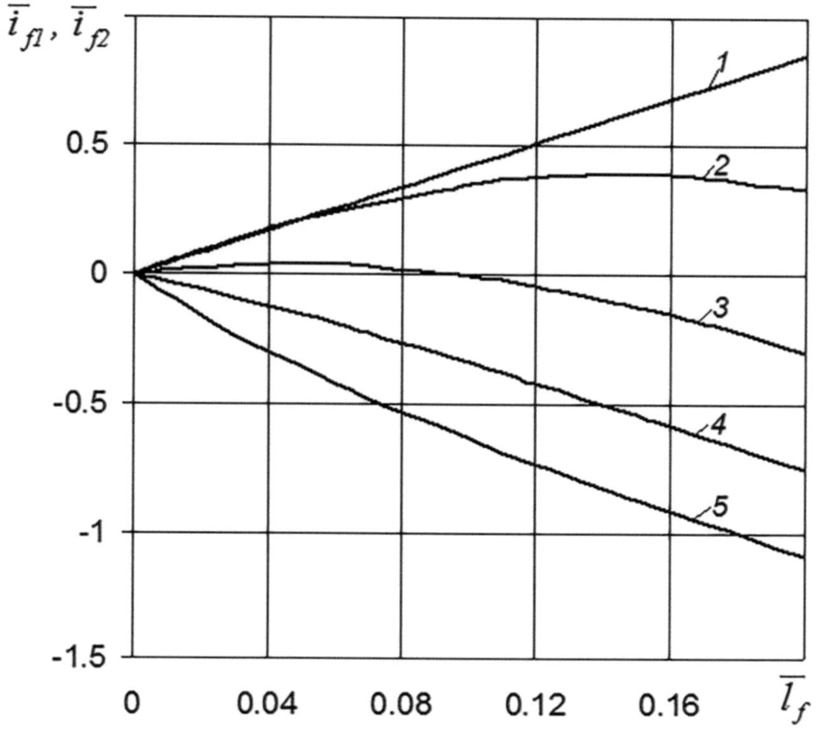

Figure 3.
Dependences: $\bar{i}_{f1}(\bar{l}_f)$, $\bar{i}_{f2}(\bar{l}_f)$: 1 – $\bar{i}_{f1}$, 2-5 – $\bar{i}_{f2}$; 2 – $\bar{S}_{20} = 0.5$, 3 – $\bar{S}_{20} = 0.25$, 4 – $\bar{S}_{20} = 0.1$, 5 – $\bar{S}_{20} = 0$.

and the ultimate thicknesses. The ratio $l_{fm} > l_{f0}$ is true for the first and here the main option. Therefore, it is real to increase the biomass when creating more comfortable conditions for it everywhere in the medium, with the possible exception of the inlet cross-section. In the second option, a strict restriction is imposed on the growth of biomass, expressed by the ratio $l_f \leq l_{fm} = 0.1$, at the inlet cross-section of the medium. It is obvious that the determination of the value l_{f0} according to Eq. (32) becomes decisive for the choice of the calculation procedure for a known value l_{fm}. **Figure 4** presents the results of the corresponding calculations for four characteristic values $\bar{\lambda}_i$. Here, a change of l_{f0} with a large margin from 0 to 0.4 was allowed. In fact, the value 0.4 is comparable to the porosity, for example, of the granular media. Thus, the value $\bar{l}_{f0}$ cannot in principle exceed the value of $\bar{l}_{fm}$, which is usually significantly less than 0.4. Nevertheless, as will be shown below, the data on $\bar{l}_{f0}$ can be useful not only for choosing a calculation variant, but also for technological calculations. A high sensitivity of $\bar{l}_{f0}$ with respect to biokinetic parameters, but especially to $\bar{k}_{di}$, is obvious from **Figure 4**. In fact, small changes in $\bar{k}_{di}$ cause incomparably large changes in $\bar{l}_{f0}$. It is natural that an increase in the rate of biomass loss leads to a thinning of the biofilm. Based on Eq. (32), it is easy to indicate such a relationship between the model coefficients, in which the biomass is not able to accumulate at all. It is described by the equation $\psi_B = \bar{S}_{20}/(\bar{S}_{20} + \bar{K}_{s2})$. On the contrary, with a decrease in $\bar{k}_{d2}$ and, thus, tending of $\bar{\lambda}_2\psi_B$ to $\tilde{\lambda}_1$ $\bar{l}_{f0}$ will grow indefinitely.

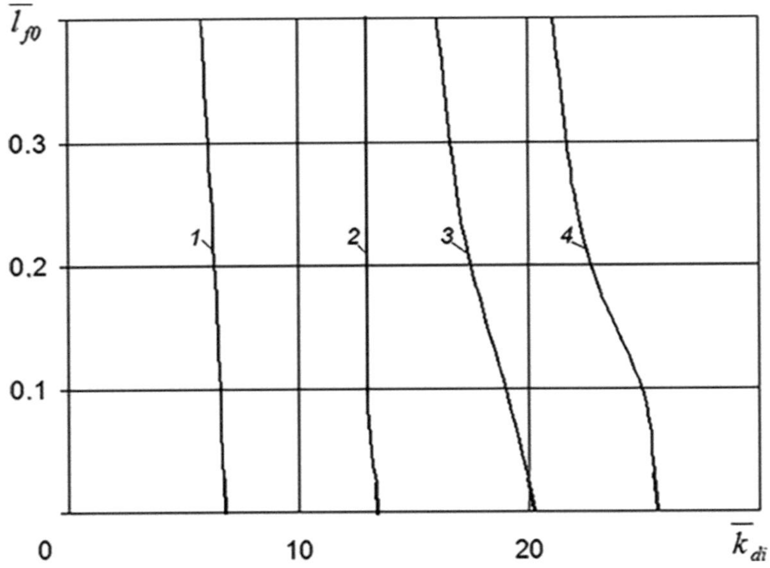

Figure 4.
Dependence $\bar{l}_{fo}(\bar{k}_{di})$: 1– $\bar{\lambda}_i = 10$, 2 – $\bar{\lambda}_i = 20$, 3 – $\bar{\lambda}_i = 30$, 4 – $\bar{\lambda}_i = 40$.

It is important to note that the biomass is distributed extremely unevenly along the section with the exception of the ultimate saturation zone $(1 \geq \bar{y} \geq \bar{y}_m)$ when the selected initial data is used. This, in particular, is evidenced by **Figure 5**, which shows the profiles calculated by (42) for $\bar{\lambda}_i = 30, k_{di} = 17$ and 18. The calculations were performed in parallel for the two above-mentioned options. It is natural that the nature of the

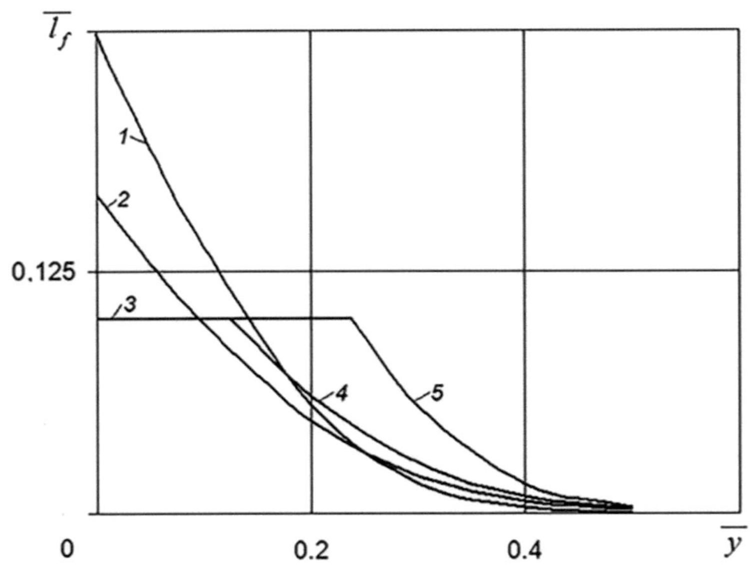

Figure 5.
Change in relative thickness of biofilms along bioreactor medium: 1, 2 – $\bar{l}_{fm} > l_{fo}$; 3-5 – $\bar{l}_{fo} > l_{fm}$; 1, 3 *u* 5 – $\bar{k}_{di} = 17$; 2, 3 *u* 4 – $\bar{k}_{di} = 18$.

biomass distribution for them differed significantly. In the case of restrained biomass growth, a significant amount of it is at a greater distance from the inlet section than in the case of unlimited growth (combined graphs 3, 4 and 3, 5). The utilization of dissolved organics should inevitably occur with less intensity within the ultimate saturation zone, but more intense outside. In order to assess the consequences of severe limitation of biomass growth for the efficiency of biofiltration process, the distributed concentrations of the initial and newly formed substrates were calculated. First of all, the distribution functions $\overline{S}_i(\overline{y})$ were found for the case $l_{fvm} > l_{f0}$ with $\overline{\lambda}_i = 20$ and three values of $\overline{k}_{di}$. The corresponding curves calculated by Eqs. (29) and (46) are shown in **Figure 6**. As noted earlier, a decrease in the rate of biomass loss contributes to its greater concentration, so that both substrates degrade more strongly throughout the entire bioreactor medium with the stable functioning of the microbiocenosis.

It is obvious that any redistribution of biomass along the bioreactor should be reflected in the appropriate way in the shape of the profiles $\overline{S}_i(\overline{y})$. **Figure 7** indicates the significance of such changes in the two options under consideration, but only within the section (bioreactor). Indeed, the pairs of the calculated curves (1 and 4, 2 and 3) corresponding to $\overline{\lambda}_i = 40$ and two values of $\overline{k}_{di}$ diverge significantly in the area $1 \geq \overline{y} \geq \overline{y}_m$, but they quickly converge with a subsequent increase in $\overline{y}$. Moreover, the calculation of the content of both the primary and the secondary substrates in the filtrate gives almost the same results with and without taking into account the restriction on biomass growth. Thus, it is permissible to determine the output concentrations of organic pollution components, ignoring the inaccessibility of a part of the pore space for the biological phase.

In conclusion, changes in the residual content of dissolved organic matter due to variation in the loss of biomass on account of the detachment, respiration and grazing were analyzed. The relative value $\Delta\overline{S}$ was finally calculated by Eq. (52) for three values of $\overline{\lambda}_i$. In this case, the range of the calculations for $\overline{k}_{di}$ was from 10 to 30. The

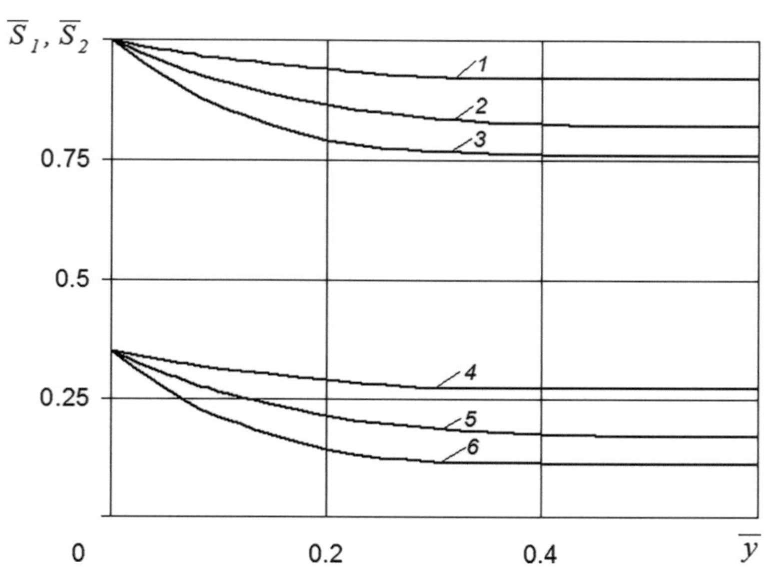

Figure 6.
Decrease in concentrations of primary and secondary substrates along bioreactor: 1-3 – $\overline{S}_1$, 4-6 – $\overline{S}_2$; 1, 4 – $\overline{k}_{di} = 13$; 2, 5 – $\overline{k}_{di} = 12$; 3, 6 – $\overline{k}_{di} = 11$.

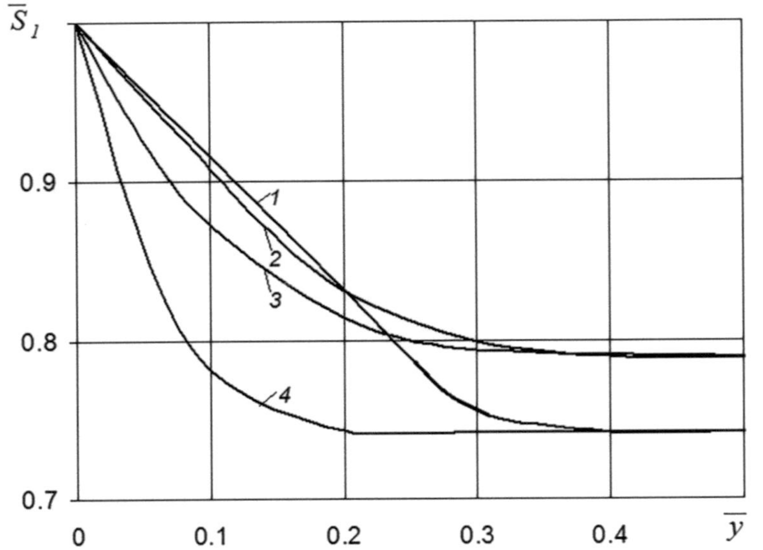

Figure 7.
Decrease in concentration of primary substrate along bioreactor medium: 1, 2 – $\overline{l}_{fo} > l_{fm}$; 3, 4 – $\overline{l}_{fo} < l_{fm}$; 1, 4 – $\overline{k}_{di} = 21$; 2, 3 – $\overline{k}_{di} = 23$.

corresponding set of graphs of the dependence $\Delta\overline{S}(\overline{k}_{di})$ is shown in **Figure 8**. Naturally, the calculated curves are limited from above by the total initial value 1.35. Once $\Delta\overline{S}$ becomes formal equal to this value it means the absence of active biomass in the section of DSMBT under consideration and, as a result, the complete incapacity of the

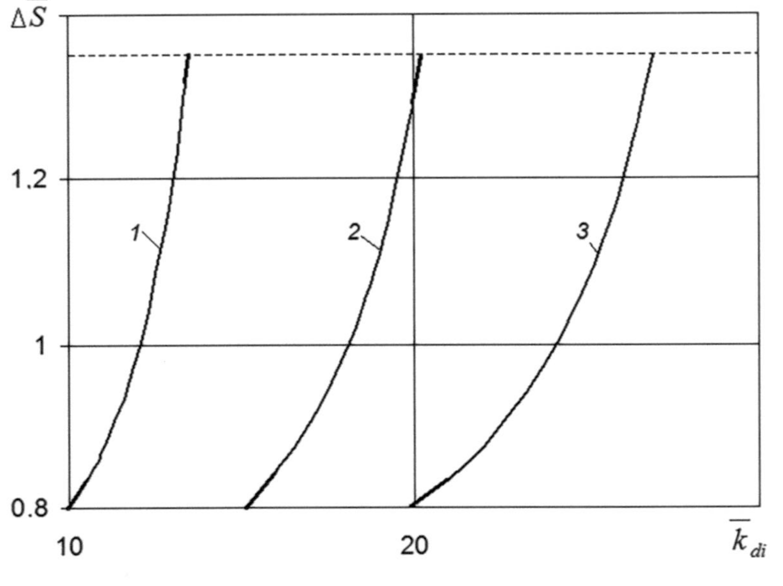

Figure 8.
Dependence $\Delta\overline{S}(\overline{k}_{di})$: 1 – $\overline{\lambda}_i = 20$, 2 – $\overline{\lambda}_i = 30$, 3 – $\overline{\lambda}_i = 40$.

latter. With a decrease in $\overline{k}_{di}$, the thickness $\overline{l}_f$ increases, verging towards $\overline{l}_{fm}$. Evidently, the best quality of water treatment here is achieved if $\overline{l}_f$ becomes equal to $\overline{l}_{fm}$ or, in other words, the entire medium will contain the maximum possible amount of active biomass. Then the corresponding maximum values of the output concentrations $\overline{S}_{1e}$ and $\overline{S}_{2e}$ are simply calculated by Eqs. (14) and (15) and as a result

$$\Delta\overline{S}_{\max} = \chi_t \overline{l}_{fm}\left(\overline{k}_{d1} + \overline{D}\overline{k}_{d2}\right). \tag{53}$$

3. Conclusions

Summing up, it should be stated that a method for engineering calculation of anaerobic biofiltration has been developed. It can be successfully applied in the case of any porous media with a regular structure. Here it is specified in relation to a biological treatment plant of a special design with a pseudo-porous medium. The method makes it possible to reliably predict the stable effect of any anaerobic biofilm, the concentration of biomass in the bioreactor medium, its permeability, and, most importantly, the concentration of organic pollution in the filtrate. It can serve as a basis for substantiating rational technological and design parameters of DSMBT. It is also easy to use the method for granular media with minimal adjustments to the calculated dependencies. This development is based on a number of assumptions that correspond to real situations with biofiltration under anaerobic conditions. Further improvement of this method is possible taking into consideration in addition:

- limitation of the decomposition rate of the primary substrate,
- inhibition of the decomposition of the secondary substrate,
- the influence of the acid–base status of the aquatic environment,
- isolation and special consideration of the stage of biooxidation,
- influence of biomass content on the rate of its loss,
- complex composition of the primary substrate.

Author details

Vadym Poliakov
Institute of Hydromechanics NAS Ukraine, Kyiv, Ukraine

*Address all correspondence to: vpoliakov.ihm@gmail.com

References

[1] Gvozdyak PI. Microbiology and biotechnology of water treatment: Quo vadis? Khimiya Tekhnologiya Vody. 1989;**11**:854-862 (in Russian)

[2] Kulikov NI, Ya NA, Omelchenko NP, Chernyshov VN. Theoretical Foundations of Water Treatment. Donezhk: Noulidzh (Donezhkoe otdelenie); 2009. p. 298 (in Russian)

[3] Sabliy LA. Physicochemical and Biological Wastewater Treatment of High-Concentrated Wastewater. Rivne: NUVGP; 2013. p. 292 (in Russian)

[4] Morgan-Sagastume J, Jiménez B, Noyola A. Anaerobic-anoxic-aerobic process with recycling and separated biomass for organic carbon and nitrogen removal from wastewater. Environmental Technology. 1994;**15**: 233-243. DOI: 10.1080/09593339409385424 (in Russian)

[5] Parameswaran P, Anderson PR. Biosolids mineralization in an anaerobic-aerobic combines reactor system. Water Research. 2007;**41**: 2739-2747. DOI: 10.1016/j.watres.2007.02.048 (in Ukrainian)

[6] Park S-M, Jun H-B, Hong S-P, Kwon J-C. Small sewage treatment system with an anaerobic-anoxic-aerobic combined biofilter. Water Science and Technology. 2003;**48**:213-220. DOI: 10.2166/wst.2004.0844 (in Russian)

[7] Sabliy L, Kuzminskiy Y, Zhukova V, Kozar M, Sobczuk H. New approaches in biological wastewater treatment aimed at removal of organic matter and nutrients. Ecological Chemistry and Engineering S. 2019;**26**(2):331-343. DOI: 10.1515/eces-2019-0023 (in Russian)

[8] Zee van der FP, Villaverde S. Combined anaerobic-aerobic treatment of ago dyes. A short review of bioreactor studies. Water Research. 2005;**39**: 1425-1440. DOI: 10.1016/j.watres.2005.03.007 (in Russian)

[9] Gvozdyak P. Based on the principle of bioconveyer (Biotechnology of environment safety). Visnyk of NAS of Ukraine. 2003;**3**:29-36. Available from: http://dspace.nbuv.gov.ua/handle/123456789/70340 (in Russian)

[10] Dmitrenko GN, Gvozdyak PI, Дмитренко ГН, Гвоздяк ПИ. Biotechnology of treatment of high-concentrated wastewater from organic solvents. Khimiya Tekhnologiya Vody. 2002;**24**:185-190 (in Russian)

[11] Endryus DP. Development of dynamical model and strategy of management for the process of anaerobic decomposition. In: Jems A, editor. Mathematical Models of Water Pollution. Moscow: Mir; 1981. pp. 321-345

[12] Buffiere P, Steyer JP, Fonade C, Moletta R. Modeling and experiments on the influence of biofilm size and mass transfer in a fluidized bed reactor for anaerobic digestion. Water Research. 1998;**32**:657-668. DOI: 10.1016/S0043-1354(97)00261-3

[13] Knobel AN, Dewis AE. A mathematical model of a high sulphate wastewater anaerobic treatment system. Water Research. 2002;**36**:257-265. DOI: 10.1016/s0043-1354(01)00209-3

[14] Haarstrick A, Mora-Naranjo N, Meima J, Hempel DC. Modeling anaerobic degradation in municipal landfills. Environmental Engineering Science. 2004;**21**:471-484. DOI: 10.1089/1092875041358494

[15] Droste RL, Kennedy KJ. Dynamic anaerobic fixed-film reactor model. Journal of Environmental Engineering. 1988;**114**:606-620. DOI: 10.1061/(ASCE)0733-9372(1988)114:3(606)

[16] Escudie R, Conte T, Steyer JP, Delgenes JP. Hydrodynamic and biokinetic models of an anaerobic fixed – bed reactor. Process Biochemistry. 2005; **40**:2311-2323. DOI: 10.1016/j.procbio.2004.09.004

[17] Aspe E, Marti MC, Roeckel M. Anaerobic treatment of fishery wastewater using a marine sediment inoculum. Water Research. 1997;**31**: 2147-2160. DOI: 10.1016/S0043-1354(97)00051-1

[18] Merkel W, Mans W, Szewzyk U. Krauth K population dynamics in anaerobic wastewater reactors modeling and in city characterization. Water Research. 1999;**33**:2392-2402. DOI: 10.1016/S0043-1354(98)00454-0

[19] Ribes J, Keesman K, Spanjers H. Modeling anaerobic biomass growth kinetics with a substrate threshold concentration. Water Research. 2004;**38**: 4502-4510. DOI: 10.1016/j.watres.2004.08.017

[20] Dmitrenko GN. Anoxic microbial processes in water treatment. Khimiya Tekhnologiya Vody. 2005;**27**: 85-103

[21] Zapolsky AK, Mishkova-Klimchenko NA, Astrelin IM, Brik MG, Gvozdyak PI, Knyazkova TV. Physicochemical Foundations of Technologies of Wastewater Treatment: Textbook. Kyiv: Libra; 2000. 552 p

[22] Banik GC, Viraraghavan T, Dague RR. Low temperature effects on anaerobic microbial kinetic parameters. Environmental Technology. 1998;**19**: 503-512. DOI: 10.1080/09593331908616706

[23] Cayless SM, Marques DMLM, Lester IN. The effect of transient loading, pH and temperature shocks on anaerobic filters and fluidised beds. Environmental Science & Technology Letters. 1989;**10**:951-968. DOI: 10.1080/09593338909384817

[24] Sánchez E, Borja R, Weiland P, Tramieso L, Martin A. Effect of temperature and pH on the kinetics of methane production, organic nitrogen and phosphorus removal in the batch anaerobic digestion process of cattle manure. Bioprocess Engineering. 2000; **22**:247-252. DOI: 10.1007/s004490050727

[25] Aguilar A, Casas C, Lema JM. Degradation of volatile fatly acids by differently enriched methanogenic cultures: Kinetics and inhibition. Water Research. 2002;**29**:505-509. DOI: 10.1016/0043-1354(94)00179-B

[26] Kus F, Wiesmann U. Degradation kinetics of acetate and propionate by immobilized anaerobic mixed cultures. Water Research. 1995;**29**:1437-1443. DOI: 10.1016/0043-1354(94)00285-F

[27] Zonta Z, Alves MM, Flotas X, Palats J. Modelling inhibitory effects of long-chain fatty acids in the anaerobic digestion process. Water Research. 2013;**47**:623-636. DOI: 10.1016/j.watres.2012.12.007

[28] Angelidaki I, Ellegaard L, Ahring BK. A comprehensive model of anaerobic bioconversion of complex substrates to biogas. Biotechnology and Bioengineering. 1999;**63**:363-372. DOI: 10.1002/(SICI)1097-0290(19990505)63:3<363::AID-BIT13>3.0.CO;2-Z

[29] Skiadas IB, Gavala HN, Lyberatos G. Modelling of the periodic anaerobic baffled reactor (PABR) based on the

retaining factor concept. Water Research. 2000;**34**:3725-3736. DOI: 10.1016/S0043-1354(00)00137-8

[30] Tartakovsky B, Guiot SR. Modeling and analysis of layered stationary anaerobic granular biofilms. Biotechnology and Bioengineering. 1997; **54**:122-130. DOI: 10.1002/(SICI)1097-0290(19970420)54:2<122::AID-BIT4>3.0.CO;2-N

[31] Poliakov VL. Modelling biofiltration of water with limited content of organic substrate. Aerobic biofilm. Dopovidi NAS of Ukraine. 2011;**5**:72-77. Available from: dspace.nbuv.gov.ua/handle/123456789/37555

[32] Poliakov VL. Modelling biofiltration of water with limited content of organic substrate Bioreactor-filter. Dopovidi NAS of Ukraine. 2011;**7**:58-66. Available from: dspace.nbuv.gov.ua/handle/123456789/38138

[33] Cakir FU, Stepstrom MK. Greenhouse gas production a comparison between aerobic and anaerobic wastewater treatment technology. Water Research. 2005;**39**: 4197-4203. DOI: 10.1016/j.watres.2005.07.042

[34] Taran M, Ozturk L. Comparative evaluation of longitudinal dispersion of liquid in non-biological and anaerobic fixed film reactors. Environmental Technology. 1997;**18**:45-53. DOI: 10.1080/09593331808616511

[35] Huang J-S, Jih C-G. Deep-biofilm kinetics of substrate utilization in anaerobic filters. Water Research. 1997; **31**:2309-2317. DOI: 10.1016/S0043-1354(97)00068-7

[36] Poliakov VL. Modeling the action of anaerobic biofilm. Report of NAS Ukraine. 2021;**6**:52-58. DOI: 10.15407/dopovidi2021.06.052

Chapter 5

Sustainability and Rural Livelihood Security Based on Biomass Gasification: An Assessment

Varinder Singh Saimbhi

Abstract

The applications of gasification products comprise impending use as process output that can be converted into mechanical, electrical and/or heat energy use in several industries and sectors including rural households. Different types of agricultural and forestry residues, energy crops, dairy-house, piggery, poultry, domestic and industrial wastes can be used as feedstock. With or without pre-treatment, the feedstock biomass can be gasified under different technological conditions viz. in a biochemical digester (biogas plant) or in a thermal digester (gasifiers), to find out what are the most suitable conditions for maximum energy outputs. Rural livelihood safety by the use of biogas plants was also assessed. The savings in fuel, overall family income, living cost, a slurry of biogas plants used as manure, and reduction of greenhouse gases at the domestic level were also studied. Overall annual family income varied from Indian Rupee (INR) 2,50,000–27,50,000. Annual livelihood cost was averaged at INR 1,66,714 and 1,83,529, respectively, with and without the usage of biogas plants. Biogas plant usage helped families save INR 10,295 (with savings of 1389 kg of fuel wood and nine cooking gas cylinders). Biogas plant usage also prevented methane (755 kg) and ammonia (267 kg) gas emissions annually.

Keywords: sustainability, rural living, gasification, biogas plants, assessment, environment

1. Introduction

Sustaining livelihood in today's world regarding quality of life, better living, economy, growth, employment and production needs energy. Shortage in the supply of energy, by any means, could disturb the national economic process and livelihood of people. India is deficient in energy. It has already been indicated that nature's energy reserves are depleting, and in the coming days, it will become difficult for us to maintain energy availability [1]. Agricultural and forestry residues, energy crops, dairy-house, piggery, poultry, domestic and industrial waste-based biomass as renewable resources will play a key part in substituting conventional fuels. About 83% of parts will be of biomass from all renewable energy generation resources [2]. Agricultural biomass is readily available and can be used to drive energy continuously. Future regional as well as rural energy needs would have to be based mainly

IntechOpen

on renewable resources, whereas implementing and integrating systems of regional energy sustainability for bigger cities, industries, towns, villages, smaller settings, etc., remains a problem even now [3]. Renewable energy sources, technologies and user applications are becoming more relevant and also call upon society to maximize and universalize their use as early as possible [4].

Farmers having abundance of agricultural wastes, a local source of energy, can become energy source supplier instead of producers of raw material. For the supply of electricity and heat at farms, biochemical conversion (anaerobic digestion) of livestock excreta would be a suitable method. Spent-digested wastes used as fertilizer virtually resulted in wider reutilizing of nutrients [2]. Gas formation through bacteria in anaerobic digestion depends upon number of environmental factors, with temperature variation being the important parameter. Bacterial activity for anaerobic digestion works best in the temperature range of 20–55°C. Anaerobic gasification of biomass at higher temperatures, within the given range, ensues swiftly. Gas production doubles at an increment of about each 5°C of temperature, below 20°C, it decreases severely and nearly halts at 10°C [5]. Therefore, during colder times when gas production from biochemical conversion (anaerobic) process retards it would be wiser to shift to the thermo-chemical conversion of biomass to balance the energy demands for sustaining rural livelihood.

Thermo-chemical conversion for gasification of biomass is the most promising route for biomass utilization [6] and is regarded as a sustainable energy technology used for waste management and producing renewable fuel [1]. Capturing atmospheric CO_2 and sulfur became a topic of interest in compatibility with conversion of biomass, environment benefits and is inexpensive over a wider capacities. Through thermal route, biomass can be converted to gaseous fuel in limited atmospheric conditions [6].

Depending upon location and owner's choice, transformation of variety of waste into usable ones may convert waste into a resource. Animal farm wastes converted into biogas make it an environment friendly reusable form of energy source [7, 8]. In agro-based country like India, vast quantities of wastes comprising cattle dung are produced endlessly. Biogas plants could convert, economically, the wastes into energy that advantageously helps in solving regional as well as rural power crisis. Biogas plants have a bright future in city corporations, metro and rural areas, producing tonnes of biological wastes, which otherwise go to waste and pollute the environment. Wastes can be utilized appropriately for power production and saving the environment from pollution [9]. Punjab state of India houses thousands of working family-size biogas plants with 5–10 cattle heads. About 4000 dairy farms with cattle head capacity ranging from 50 to 500 produce milk every day. Also, a large number of stray cattle are being housed in public cattle yards ('*Goushalas*'). A large quantity of biogas can be produced using large capacity (50–500 m^3 per day) biogas plants from the huge availability of cattle dung [10].

Under anaerobic conditions, biodegradable material is acted upon by methanogenic bacteria to produce biogas [11, 12]. Biogas contains 50–70% methane (CH_4), 30–40% carbon dioxide (CO_2), apart from other gases like hydrogen (H_2), ammonia (NH_3), hydrogen sulphide (H_2S), nitrogen (N_2) and water vapors (H_2O). Methane and hydrogen in biogas are energy sources that can be easily converted into heat energy (for cooking and lighting) and mechanical energy via internal combustion/gas engines. Additionally, biogas spent slurry can be used as organic fertilizer in agricultural farms to cut down usage and dependency on chemical fertilizers. In this way, biogas plants helps replace conventional energy sources, reducing emissions of harmful gases and mitigating the bad effects of carbon released to environment [13]. Produced biogas can be used locally or may transported over a longer distance by upgrading to natural gas equivalent quality [8].

Biogas plant economy depends on noteworthy investment costs, operation and maintenance costs, raw materials (available freely, most of the time), like cattle/animal dung, poultry droppings, organic wastes of household and industries, sewage sludge, aquatic and terrestrial plants etc., costs and income gained from the sale of gas and dried slurry. Supplementary value additions may be enriched biofertilizer production, greenhouse gas emission reduction and decreased cooking time and fuelwood collection. The system economy is also site-specific as prevalent market rates affect the cost of input/output, farming practices followed and management practices adopted by the involved community, etc. [14]. Certain factors, such as the incorporation of advanced technologies of feedstock processing, energy conversion and knowledge of complexities between plant size and costs involved, affect the impending biogas energy cost [15].

It has been well illustrated that biogas production and use have the capabilities to reduce the fuelwood usage and deforestation, cessation of indoor air pollution, thereby improving women's and children's health, reduction of greenhouse gases to mitigate global warming, availability of organic fertilizer for improving soil health, generating income, better sanitary conditions, reducing dependence on imported fuels etc. [16, 17]. The availability of enhanced livelihood facilities to users has been highlighted despite the fact that the user's awareness and appreciation were partial. The need for householders training, in particular educating the women, was proposed for adequateness of feedstock, system upkeep, environmental benefits and possible livelihood security [18].

Rural livelihood hangs on agriculture for sustainability. Without access to electricity, firewood has been used as the basic cooking fuel. The present condition has an adverse effect on the environment and is unsustainable. The biogasification addresses the said apprehensions as being the simple yet powerful option for providing better fuel for cooking and lightning, empowering households by saving time with lesser environmental concerns like dispensation of animal wastes, contamination of groundwater, greenhouse gas emission and threat of climate change [19]. Households, using biogas from 18 to 24 months were 41.4%, having large size (>20 m^3) biogas plants were 69.4%, have incurred 50–60 thousand Pak Rupee (PKR) as making cost were 41.7%, out of a survey sample. Monthly savings of 3–3.5 thousand PKR by 30.6% and more than 3.5 thousand PKR by 20.8% were made by households using biogas in comparison to conventional fuel costs [20].

Focus on the use of biogas as a renewable energy technique to realize socioeconomic and environmental sustainability was made that combines the production, consumption and natural/industrial ecosystems research. A framework was developed for adopting biogas energy in industrial and rural ecosystems as bottom-of-the-pyramid. It was suggested to embed a meta-dimension into the dimensions of biogas industrial ecosystem to enable socioeconomic and environmental sustainability [8]. In developing countries and humanitarian camps, organic waste management addresses hygiene and sanitation, which otherwise causes serious health issues that may lead to premature deaths [21]. Net Present Value (NPV), Internal Rate of Return (IRR), Payback Period (PBP) and Net Benefit Increase (NBI) of biogas plant's impact on livelihood were studied in Bangladesh. Respondents belonging to 30- to 50-year age group were 54%, those belonging to the business group were 24%, up to higher secondary education were 36%, those having an average family size of 4 (3.78) were

36%, those belonging to middle-income (1,00,001–1,50,000 Bangladesh Taka, Tk.) group were 8%, and those belonging to high-income (>1,50,000 Tk.) group were 92%, out of the surveyed sample. Increase in knowledge, skill, work capacity, nutrition, health and education were found in 58, 64, 70, 78 and 66%, and increase in leadership, women liberation, common rules, mutual support, networking connections were found in 54, 60, 52, 64 and 62% with the use of biogas [11].

A model was developed for making detailed inventories of producing electricity from biogas. Types of biogas systems studied were organic waste landfill-based and dairy cattle waste-based. Life cycle assessment (LCA) showed different sub-stages of the systems. It was concluded that LCA enables producers, decision-makers and government agencies to recognize and improve opportunity areas of the technology [12].

Convenient usage of both biochemical (anaerobic digestion) and thermo-chemical conversion technologies, the energy production at farms can become surplus by using the total biomass produced. Agricultural economy may recover and sustainable regional development may become feasible. A combination of renewable technologies in energy supply for rural livelihood sustainability has a promising potential.

With its numerous rewards for rural households empowering and making their livelihood sustainable, a review discussion was made on the research work in gasification of biomass for recommending the sustainable solution with technologies, viz. biochemical and thermal gasification and their costs when accounting for certain conditions; viz. family size, livelihood cost among others along with users/stakeholders interests. An assessment was made on the socioeconomic effects of biogas plants on the livelihood security of rural households in the selected villages of Punjab State of India.

2. Biochemical gasification

In a country like India, where about 68.8% of the total population lives in rural areas [22], one of the alternative energy source is biogas. Biogas, a product of anaerobic digestion in the absence of air of cellulosic biomass, like cattle dung, poultry droppings, pig excreta, human excreta, crop residues etc., abundantly available in rural areas, is a suitable fuel for providing heat and operating stationary engines. This anaerobic digestion results in the production biogas, a mixture of combustible gases, mostly containing 50–60% methane, 30–40% carbon dioxide, 1–5% hydrogen and traces of nitrogen, hydrogen sulphide, oxygen, water vapors, etc. [23]. Anaerobic digestion not only provides valuable fuels and enhances the fertilizer value of the waste but also provides a conventional, safe, aesthetical and economical waste disposal method. The biogas plant design is unswervingly linked to its hydraulic retention time (HRT). The HRT is the time span (days) for which a mixture of water and cattle dung remains inside an enclosed digester for gas generation and after HRT, its biological capability is diminished. The HRTs of biogas plants differ based on their location in India. Most of the plains in India, including Punjab, have 40 days of HRT.

2.1 Design models of biogas plant

Biogas plants are mostly categorized as batch type, and continuous type. Batch-type biogas plants are appropriate where daily supplies of raw waste materials are difficult to obtain and are most suitable for digestion of crop biomass. In continuous-type biogas plants, supply of gas is continuous, and the digester is fed with biomass regularly. These types of biogas plants may be single-stage, double or multiple-stage. There are two basic continuous type biogas plants designs that are popular in India; viz., floating drum type, and fixed dome type. Popular models in floating drum type are '*Khadi*' and Village Industry Commission (KVIC) model, whereas, '*Janta*' model, and '*Deenbandhu*' model are popular among fixed dome types [5].

2.2 Selection of size of biogas plant

The size or capacity is the quantity of biogas produced in cubic metres (m^3) on 24 24-hour basis. Quantity of cattle dung available or number of family members in a household quantifies the biogas plant capacity. As a thumb rule, 1 kg of cattle dung can produce approximately 0.04 m^3 of biogas, or 0.34–0.42 m^3 of biogas is required per person for cooking food. Either of these criteria can be used for finding the other. Therefore, it is presumed that, as the biological process sets in, 1 m^3 of biogas is produced from 25 kg of cattle dung. As a matter of fact, as ordinary cattle, depending on a number of factors, produce about 10–20 kg of dung [23]. The plant size calculations, based on the above data, are shown in **Table 1**.

2.3 Cost of installation of biogas plants

The costs of civil construction including the cost of steel required for each type of biogas plants was taken as per criteria adopted [4]. The cost of installation of different family-size biogas plants that includes material and labour costs at the prevalent market rates is given in **Table 2**.

The Indian government provides a fixed amount of financial assistance, for promoting the use of biogas, through the state energy development agency (PEDA, Punjab Energy Development Agency in case of Punjab state) in Indian rupees, as given in **Table 3** [24].

Sr. no.	Capacity of biogas plant (m^3)	No. of animals required	Quantity of dung required (kg)	Cooking for number of persons
1	2	3–4	50	4–5
2	3	5–6	75	7–8
3	4	7–8	100	10–11
4	6	10–12	150	4–16

Table 1.
Number of persons served, requirement of dung and number of animals for different sizes of biogas plants.

Sr. no.	Biogas plant models	Plant capacity			
		2 m^3	3 m^3	4 m^3	6 m^3
1	KVIC type	30,000	37,000	43,000	55,000
2	'*Janta*' type	26,000	30,000	32,000	40,000
3	'*Deenbandhu*' type	20,000	25,000	30,000	35,000

Table 2.
Installation costs of different types of biogas plants.

Sr. no.	Regions and beneficiary category	Capacity of biogas plants in cubic metre/day (INR per plant)					
		1	2–4	6	8–10	15	20–25
1	(a) Hilly/North Eastern Region States (b) Island; and (c) Scheduled Castes (SC)/Scheduled Tribes (ST)	17,000	22,000	29,250	34,000	63,250	70,400
2	All other States and Categories	9800	14,350	22,750	23,000	37,950	52,800
3	Additional fixed Subsidy for						
	(i) Biogas plant if linked with sanitary toilet	1600	1600	1600	1600	NA	NA
	(ii) Biogas plant if linked with approved Biogas slurry filter unit	1600	1600	1600	1600	1600	1600
4	Turnkey job fee for construction, supervision, commissioning and free operation and maintenance warranty for 5 years of trouble-free operations	INR 3000 per plant for 1–10 m^3 and INR 5000 per plant for 15–25 m^3 size. This turnkey job fee is applicable only for plants involving onsite construction such as fixed dome design '*Deenbandhu*' Model, floating gasholder. KVIC Model. Turnkey job fees will not be eligible for prefabricated/manufactured biogas plants.					

NA: Not applicable.

Table 3.
Central financial assistance (CFA) under the biogas programme.

3. Thermo-chemical gasification

Thermal gasification includes incomplete burning (oxidation in limited air amount/oxidant) and reduction processes of biomass to produce combustible mixture of gases known as producer gas. In a characteristic incineration procedure usually oxygen is in excess, whereas in gasification route fuel is in excess. The combustion yields, mainly carbon dioxide, water vapor, nitrogen, carbon monoxide, hydrogen, etc., and these are passed through the burning layer of charcoal for the reduction process to occur. During this stage, both carbon dioxide (CO_2) and water vapor oxidize the char to form carbon monoxide (CO), hydrogen (H_2) and methane (CH_4). A typical composition of the gas obtained from wood gasification on volumetric basis is as; CO is 18–22%, H_2 is 13–19%, CH_4 is 1–5%, heavy hydrocarbons is 0.2–0.4%, CO_2 is 9–12%, nitrogen (N_2) is 45–55% and water vapor at 4%. Gasifiers are broadly classified into (i) updraft, (ii) downdraft and (iii) fluidized types [5].

3.1 Selection of gasifiers

An extensive review of gasifier manufacturers identified 50 producers proposing 'commercial' gasification plants; out of that, 75% were in the downdraft category, 20% were in the fluidized bed category, 2.5% were in the updraft category and

Sr. no.	Capacity (kW)	Application	Fuel and its size (mm)
1	5	Water pumping	Rice Husk, As such
2	10	Water pumping or electricity generation	Wood chips, maize cobs, cotton and pulses sticks, 50
3	25	Electricity generation	Wood chips, maize cobs, cotton and pulses sticks, 100
4	40	Electricity generation	Wood chips, maize cobs, cotton and pulses sticks, 120
5	60	Electricity generation	Wood chips, maize cobs, cotton and pulses sticks, 150

Moisture content of fuel: Less than 15%; Engine de-rating: 15–25%; Diesel replacement: 70–75%; and Fuel consumption: 1–1.3 kg/kWh.

Table 4.
Performance of biomass gasifier-diesel engine system.*

Sr. no.	Items	Pattern of assistance (INR)
1.	Biomass gasifier for captive power applications in industries and other institutions	Electrical – INR 2500 per kWe with dual fuel engines
2.	Distributed/off-grid power for villages and up to 2 MW Grid-connected power projects	Electrical – INR 15,000 per kWe with 100% gas engines
3.	Biomass gasifier for captive thermal applications in industries and other institutions	Thermal – INR 2.0 lakh per 300 kWth for thermal use

Table 5.
The pattern of central financial assistance for biomass gasifiers.

2.5% were of different other categories [25]. While very small facts were provided on authentic hours of working experience, gasifier's turn-down ratios, efficiencies, emissions and price features. In all, none of the gasifier producers supplied the full guarantee for the practical working of their technology [26].

For meeting the thermal and electrical requirements of rural households, downdraft gasifiers have been developed in the range of 5–60 kW, as given in **Table 4**. A 5 kW gasifier will be suitable for a family of 4–6 persons. After cleaning producer gas through filters, diesel engine gen-sets can be operated, and up to 70% of diesel savings can be made. Annual savings of up to INR 20,000 can be achieved by operating the system for 1500 hours, and one can recover the cost of the gasification system in 3 years [27].

The central financial assistance (CFA) is provided by the Government of India through the state energy development agency PEDA in the form of back ended subsidy for installation of waste-to-energy biomass gasifier projects for recovery of energy from urban, industrial, agricultural waste/residues and municipal solid wastes. The details of the fixed amount of financial assistance given in the biomass gasifier programme are given in **Table 5** [28].

4. Technology specific barriers

Gasification of biomass, as discussed, is intricate and diverges by several means, for example, feed properties desirable (i.e. farming/forestry biomass, cattle dung, etc.), span of lifetime (long, medium and short-term), natures of usage (cooking, thermal, etc.), and upkeep required (monthly, weekly and daily). Discrepancies in the type of biomass gasification know-how, practical performance doubts, market inadequacies such as controlled energy sector, high investment and transaction costs, restricted access, lack of information and competition, trade barriers etc., economic and financial reasons such as non-viable, need for incentives, long payback periods, small market size, capital costs, lack of access to capital, credit, financial institutions, high up-front cost to investors etc. Institutional barriers such as deficiency of organizations to propagate information, permitted/monitoring outline, difficulties in realizing monetary inducements, stakeholder's interest and choice building, clash of welfares, deficiency of research and development, private-sector involvement, specialized establishments, etc. Procedural barriers such as the absence of typical codes and documentation, expert workforces, training services, operation and maintenance services, businesspersons, structural limitations, unreliable products, etc. Societal obstacles include the absence of customer recognition and social acceptance, and other incidental barriers such as uncertain government policies, high-risk perception and environmental conditions. However, policies and programmes initiated by the government have made an attempt to address some of these barriers to propagating the technology to rural masses. The policy opportunities in overpowering such obstacles for the advancement comprise research and development aimed at reducing price, consistent working, investment grants, extensive demonstrations, work-based monetary inducements, reasonable charges instead of biomass power, performance assurances, formation of a big linkage of businesspersons and trained individuals for the manufacture, set up, upkeep of skill structures, training and awareness are the main concern [23, 25].

5. Assessment methodology

A review was piloted in designated villages/communities of district Patiala of Punjab state of India. A feedback form was prepared for the collection of essential data associated with farmer land holding, biogas plant set up or not, its kind, capacity, year of installation, cattle heads, fuel type used in the household kitchen, biogas usage and other purposes. The socioeconomic contemplations like savings in terms of replacing conventional fuel by biogas were also included in survey proforma. The data was collected by making personal contacts with the individual farmers in selected villages.

To understand the socioeconomic characteristics of domestic level use of biogas plants, the questionnaire was set with objectives of knowing the household income from different sources (like agricultural land contracts, business, private or government service and others, if any). Data on cultivatable area where slurry was used as manure, installation cost of biogas plants and their capacity, and any subsidy if given by the government to install the biogas plant were also noted. The data collection was also done to know the difference between families' livelihoods: those having biogas plants and others with no biogas plants. The survey also focused on amount of savings made by people who owned the biogas plant by means of conventional fuel savings (i.e. fuel wood, liquefied petroleum (LPG) gas cylinders and dung cakes) made with the usage of biogas in kitchen.

The influences of the use of biogas at domestic level were considered in terms of by-product (spent slurry) use as manure with decrease, if any, in the applicable dose of chemical fertilizers noted in survey proforma for calculating the savings.

The greenhouse gas emission reduction by usage of cattle dung, which is otherwise dumped in open, was also calculated on a yearly basis as described in [29, 30]. The amount of methane and ammonia gases released in kg/year is estimated by multiplying number of cattle by 112 for kg of methane and adding with multiple of number of cattle to 40 for kg of ammonia gas release estimation.

6. Assessment results and discussion

Data was collected from 45 farming households in 15 different communities of district Patiala. The communities covered in the study were Naraingarh, Gulhar, Sagra, Kullar, Noorpur, Majri Palak, Manjoli, Bhedpura, Tuga, Namada, Jagatpura, Bhima Khedi, Fatehgarh, Jafarpur and Bakhshi Wala.

6.1 Family details

The detail of family members of different households that had biogas plants and others that did not have or had non-working biogas plants is given in **Table 6**.

About 31 farmers had biogas plants for domestic use, out of which 28 plants were found in working conditions and three plants were not working. The average number of members in the family of surveyed farmers were six for both households having and not having biogas plants. All farmers under the survey work had installed '*Deenbandhu*' type biogas plant of 6 m^3 capacity. All farmers were using the biogas in their household kitchens for cooking purposes. Farmers were also disposing of the surplus cattle dung in pits where biogas plant slurry had been kept. About 12 farmers were also making dung cakes. The surveyed households had similar human resources, most of which were of medium family size.

6.2 Socioeconomic considerations

The surveyed household's cultivable land holdings, overall income and earnings from farming are given in **Tables 7–9**.

The average land holding was 12.04 acres and 8.12 acres among households having and not having biogas plants, respectively. At the same time, maximum percentage of farmers belong to semi-medium category in both groups. The households having

Family size	Households having biogas plants		Households not having biogas plants	
	Numbers	Percentage	Numbers	Percentage
Small (<5)	7	25.0	1	5.9
Medium (5–7)	16	57.1	12	70.6
Large (>7)	5	17.9	4	23.5
Total	28	100.0	17	100.0
Maximum	10		11	
Minimum	3		3	
Average	6		6	

Table 6.
Detail of family members of surveyed farmers.

Land holdings (acre)	Households having biogas plants		Households not having biogas plants	
	Numbers	Percentage	Numbers	Percentage
Marginal (<1)	2	07.2	2	11.8
Small (1–2)	3	10.7	1	05.9
Semi-med (2–4)	11	39.3	10	58.8
Medium (4–10)	9	32.1	3	17.6
Large (>10)	3	10.7	1	05.9
Total	28	100.0	17	100.0
Maximum	55		26	
Minimum	2		2	
Average	12.04		8.12	

Table 7.
Land holdings of surveyed farmers.

Income range (Rs.)	Households having biogas plants		Households not having biogas plants	
	Numbers	Percentage	Numbers	Percentage
Less than 2 lakh	0	0.0	0	0.0
2–5 lakh	7	25.0	7	41.2
5–10 lakh	15	53.6	9	52.9
10–15 lakh	3	10.7	0	0.0
More than 15 lakh	3	10.7	1	5.9
Total	28	100.0	17	100.0
Maximum	27,50,000		21,00,000	
Minimum	2,50,000		2,50,000	
Average	8,32,143		6,11,176	

Table 8.
Details of overall income from all sources of surveyed farmers.

biogas plants were holding a comparatively larger farm area as compared to households not having biogas plants.

The average earnings from all sources of households having biogas plants was around INR 8,32,143, and of households not having biogas plants was around INR 6,11,175. Most of the households fall in INR 5–10 Lakh annual income category among both the groups. Whereas the average earnings from farmland of households having biogas plants was around INR 5,85,714, and of households not having biogas plants was around INR 4,05,882. Most of the households fall in INR 2–5 Lakh annual agricultural income category among both the groups. The households having biogas plants were earning comparatively better from farms and other works as compared to households not having biogas plants.

Details of other remunerative works and income generation among the surveyed farmers are given in **Tables 10** and **11**.

Income range (Rs.)	Households having biogas plants		Households not having biogas plants	
	Numbers	Percentage	Numbers	Percentage
Less than 2 lakh	4	14.3	2	11.8
2–5 lakh	13	46.5	11	64.7
5–10 lakh	6	21.4	3	17.6
10–15 lakh	3	10.7	1	5.9
More than 15 lakh	2	7.1	0	0.0
Total	28	100.0	17	100.0
Maximum	27,50,000		13,00,000	
Minimum	1,00,000		1,00,000	
Average	5,85,714		4,05,882	

Table 9.
Details of income from farm land of surveyed farmers.

Other works	Households having biogas plants		Households not having biogas plants	
	Numbers	Percentage	Numbers	Percentage
Service	9	56.3	6	66.7
Business	6	37.5	3	33.3
Others	1	6.3	0	0.0
Total	16	100.0	9	100.0

Table 10.
Detail other remunerative works done by the surveyed households.

Other works	Income of households having biogas plants (Rs.)			Income of households not having biogas plants (Rs.)		
	Maximum	Minimum	Average	Maximum	Minimum	Average
Service	10,00,000	1,00,000	4,94,444	8,00,000	1,50,000	3,90,000
Business	7,00,000	1,50,000	3,91,667	5,00,000	3,00,000	3,83,333

Table 11.
Income earned from other works by the surveyed households.

About 16 households having biogas plants and nine households without biogas plants were engaged in government/private service, business and other works. The households with biogas plants were more business-oriented, whereas those without biogas plants were more service-oriented.

The average income earned from service and business of households with biogas plants was INR 4,94,444 and 3,91,667, respectively. as compared to INR 3,90,000 and 3,83,333 for households without biogas plants. The households having biogas plants were comparatively earning better from other works than households without biogas plants.

The details of monthly and yearly livelihood costs and annual conventional fuel (wood, cooking ga and dung cakes) savings made using biogas by the surveyed farming household are given in **Table 12**.

Livelihood cost (Rs.)	Households having biogas plants		Households not having biogas plants	
	Numbers	Percentage	Numbers	Percentage
Less than 1 lakh	4	14.3	0	0.0
1–2 lakh	14	50.0	11	64.7
2–3 lakh	10	35.7	6	35.3
More than 3 lakh	0	0.0	0	0.0
Total	28	100.0	17	100.0
Maximum	3,00,000		3,00,000	
Minimum	60,000		1,08,000	
Average	1,66,714		1,83,529	

Table 12.
The total cost of livelihood of the surveyed households.

The average annual cost of livelihood of households with biogas plants was INR 1,66,714. In contrast, it was found that the average annual cost of livelihood of farmers who did not have biogas plants at their houses was INR 1,83,529. The farming households with biogas plants were less spendthrift as compared to households without biogas plants.

6.3 Impact of biogas plants on livelihood of rural people

The details of annual savings made by using biogas in kitchens in terms of fuel wood, cooking gas cylinders and dung cakes in different households with biogas plants are given in **Table 13**.

The average annual fuel savings made by the surveyed families were about 1389 kg of wood and about nine cooking gas cylinders. Five farming households were also saving about 440 kg of dung cakes. The average annual fuel savings by the surveyed families was INR 10,295.

The farming households with biogas plants also used dried biogas plant slurry as manure, and farming households without biogas plants used cattle dung as farm yard manure. The details of total chemical fertilizer (Urea) savings made and paddy yield (in quintals, q) gained by the farming households are given in **Tables 14** and **15**.

Surveyed farmers used dried slurry as farm yard manure (FYM) on 4.7 acres, on average. The parameters taken into consideration to convert the quantity of fertilizers saved and increase in production into money with the rate of urea was Rs.2.5 per kg and rate of the paddy was INR 2040 per quintal (average rate of course and fine varieties). The average increase in yield by using dried biogas plant slurry as manure was 1.8 quintal per acre, and 48.0 kg of fertilizer was saved per acre. The average money saved by the use of biogas plant slurry was equal to INR 3957 per acre.

With the use of cattle dung as FYM alone, chemical fertilizers (Urea) savings were also made on about 4.4 acres, on average, by farming households without biogas plants. It was also found that average increase in yield by using FYM was 2.1 quintal per acre, and 57.1 kg of fertilizer was saved per acre. The average money saved by the usage of cattle dung as FYM alone was found to be equal to INR 4336 per acre.

The use of biogas plant slurry as manure was not found to be better regarding urea fertilizer savings and crop yield gain than cattle dung use as FYM in paddy crop cultivation.

Details	Fuel wood saving (kg)	LPG cylinders (No.)	Dung cakes (kg)	Total savings (Rs.)
Maximum	3000	12	600	15,450
Minimum	500	5	200	4200
Average	1389	9	440	10,295

Table 13.
Annual savings made by using biogas in kitchens.

Details	Cultivation area (acre)	Urea fertilizer saving (kg/acre)	Yield gain (q/acre)	Total benefits (Rs.)
Maximum	10	65	2.5	52,120
Minimum	2	30	1	4330
Average	4.7	48.0	1.8	18,598

Table 14.
Savings from chemical fertilizers and yield increase by using biogas plant slurry as manure in paddy cultivation.

Details	Cultivation area (acre)	Urea fertilizer saving (kg/acre)	Yield gain (q/acre)	Total benefits (Rs./acre)
Maximum	8	70	3	33,967
Minimum	1	45	1.3	5275
Average	4.4	57.1	2.1	19,078

Table 15.
Savings from chemical fertilizers and increase in crop yield by using cattle dung as manure in paddy cultivation.

6.4 Impact of biogas plants on environment protection

The details of cattle head and quantity of greenhouse gas (methane and ammonia) emission reduction by the farming households with biogas plants are given in **Table 16**.

With the use of biogas plants, the farmers prevented the emission of methane and ammonia from the open disposal of cattle dung. The average amount of ammonia and methane emissions contained yearly were about 267 and 755 kg, respectively. Whereas, with the open disposal of cattle dung by household not having biogas plants, the average

Details	Emission reduction by biogas plants (kg/year)		Emission by open disposal of cattle dung (kg/year)	
	Ammonia	Methane	Ammonia	Methane
Maximum	520	1456	320	896
Minimum	120	336	40	112
Average	267	755	176	494

Table 16.
Details of greenhouse gas emission reduction by biogas plants and emissions from open disposal of cattle dung by the surveyed households.

yearly ammonia and methane emissions were about 176 and 494 kg, respectively. This means biogas plant helps greatly in protecting the environment from greenhouse gases.

7. Conclusions

Biogas plants were successful in the outer peripheries of villages or in fields. Biogas can be burnt for cooking or lighting the house. It can also be used to run internal combustion/gas engines to generate motive power or generate electricity. Two types of economic benefits can be taken, i.e., one is it saves the energy cost to be purchased, and the other is it can earn extra money by selling to the neighbors. Owing to the problems of land availability and provisions of required feedstock, biogas plants are less successful in the interiors of communities.

Thermal gasification of biomass is an encouraging technology to replace the usage of conventional fuels and to decrease fossil CO_2 release into the atmosphere. A great potential exists with this type of renewable energy: it can consume extensive kinds of materials as feed input for energy production. Also, it can produce numerous chemicals and fuels. Abundant quantities of crop/forestry-based biomass are available, and it can be optimally used for the thermal and power requirements of communities by empowering co-operatives with the technical know-how of the technology along with convincing incentives that may change the overall energy scenario at the rural level.

The assessment study showed that the average family members of surveyed household were 6, each and the average cultivable land was 12.04 acres and 8.12 acres among households having and not having biogas plants, respectively. Out of the 45 households surveyed, 13 have income from government/private service and 7 farmers were running some business. The income of the surveyed households varied from INR 2,50,000 to Rs.27,50,000. The average annual fuel savings made by the surveyed families were about 1389 kg of wood, and about nine cooking gas cylinders. Five farming households saved about 440 kg of dung cakes. The average annual fuel savings by the surveyed families was Rs.10,295. The average annual cost of livelihood of the farmers who have and did not have biogas plants at their houses was INR 1,66,714 and 1,83,529, respectively. The farmers, having biogas plants, used dried slurry as FYM on about 4.7 acres, and the average increase in yield was 1.8 quintals per acre, and 48 kg of chemical fertilizer (Urea) was also saved per acre. The average money saved by the use of biogas plant slurry as FYM was about INR 3957 per acre. The farming households without biogas plants also saved chemical fertilizers (Urea) by using cattle dung as FYM on about 4.4 acres. The average increase in yield by using FYM was 2.1 quintal per acre, and 57.1 kg of chemical fertilizer was saved per acre. The money saved was around INR 4336 per acre. The average number of cattle heads owned by the surveyed farmers was 7. The biogas plant farmers prevented the emission of methane and ammonia to the tune of about 755 kg of methane and 267 kg of ammonia on an average per year. The average amount of ammonia and methane released to the atmosphere, by open disposal of cattle dung were about 176 and 494 kg, respectively, of households not having biogas plants.

Acknowledgements

The author duly acknowledges Er. Jatinder Singh for providing basic information regarding the use of biogas plants at farmer's level that was collected during the course of his graduate project work.

Conflict of interest

The authors declare no conflict of interest.

Author details

Varinder Singh Saimbhi
Punjab Agricultural University, Ludhiana, India

*Address all correspondence to: saimbhivs@rediffmail.com

References

[1] Adams PWR, Mcmanus MC. Small-scale biomass gasification CHP utilisation in industry: Energy and environmental evaluation. Sustainable Energy Technologies and Assessments. 2014;**6**:129-140. DOI: 10.1016/j.seta.2014.02.002

[2] Karpenstein-Machan M. Sustainable cultivation concepts for domestic energy production from biomass. Critical Reviews in Plant Sciences. 2001;**20**(1):1-14. DOI: 10.1016/S0735-2689(01)80010-1

[3] Maier S, Gemenetzi A. Optimal renewable energy systems for industries in rural regions. Energy, Sustainability and Society. 2014;**4**:9. DOI: 10.1186/2192-0567-4-9

[4] Singh KJ, Sooch S. Comparative study of economics of different models of family size biogas plants for state of Punjab. India. Energy Conversion and Management. 2004;**45**(9-10):1329-1341. DOI: 10.1016/j.enconman.2003.09.018

[5] Khan BH. Non-conventional Energy Resources. 2nd ed. New Delhi: Tata McGraw-Hill; 2006. p. 455

[6] López-González D, Fernandez-Lopez M, Valverde JL, Sanchez-Silva L. Gasification of lignocellulosic biomass char obtained from pyrolysis: Kinetic and evolved gas analyses. Proceedings of the ICE—Energy. 2014;**2014**(71):456-467. DOI: 10.1016/j.energy.2014.04.105

[7] Okello S, Pindozzi S, Faugna S, Lorenzo B. Development of bioenergy technologies in Uganda: A review of progress. Renewable and Sustainable Energy Reviews. 2013;**18**:55-63. DOI: 10.1016/j.rser.2012.10.004

[8] Surie G. Achieving sustainability: Insights from biogas ecosystems in India. Agriculture. 2017;**7**(2):15. DOI: 10.3390/agriculture7020015

[9] Balat M, Balat H. Biogas as a renewable energy source—A review. Energy Sources, Part A Recovery, Utilization and Environmental Effects. 2009;**31**(14):1280-1293. DOI: 10.1080/15567030802089565

[10] Sooch SS. Infusion of biogas technology in Punjab: A case of large capacity biogas plants. Indian Journal of Economics and Development. 2015;**11**(1):239-243. DOI: 10.5958/2322-0430.2015.00026.8

[11] Haque MA, Rahaman J. Power crisis and solution in Bangladesh. Bangladesh Journal of Scientific and Industrial Research. 2010;**45**(1):155-162. DOI: 10.3329/bjsir.v45i2.5714

[12] Reynoso EAH, Aguilar HAL, Gómez JA, Méndez MGG, Hernández AP. Biogas power energy production from a life cycle thinking. In: New Frontiers on Life Cycle Assessment - Theory and Application. IntechOpen; 2019. pp. 18-36. DOI: 10.5772/intechopen.78248

[13] Jin H, Chang Z, Ye X, Ma Y, Zhu J. Physical and chemical characteristics of anaerobically digested slurry from large-scale biogas project in Jiangsu Province. Transactions of the Chinese Society of Agricultural Engineering. 2011;**27**:291-296. DOI: 10.3969/j.issn.1002-6819.2011.01.047

[14] Taleghani G, Kia AS. Technical–economical analysis of the Saveh biogas power plant. Renewable Energy. 2005;**30**:441-446. DOI: 10.1016/j.renene.2004.06.004

[15] Singh PP, Ghuman BS, Grewal NS. Computer model for performance

prediction and optimisation of unheated biogas plant. Energy Conversion and Management. 1998;**39**:51-63. DOI: 10.1016/S0196-8904(96)00177-X

[16] Smith K, Samet J, Romieu I, Bruce N. Indoor air pollution in developing countries and acute lower respiratory tracts infections in children. Thorax. 2012;**55**:518-532. DOI: 10.1136/thorax.55.6.518

[17] Lantz M, Svensson L, Björnsson L, Börjesson P. The prospects for an expansion of biogas systems in Sweden-incentives, barriers and potentials. Energy Policy. 2007;**35**:1830-1843. DOI: 10.1016/j.enpol.2006.05.017

[18] Raha D, Mahanta P, Clarke ML. The implementation of decentralized biogas plants in Assam, NE India: The impact and effectiveness of the National Biogas and FYM Management Programme. Energy Policy. 2014;**68**:80-91. DOI: 10.1016/j.enpol.2013.12.048

[19] Saidmamatov O, Rudenko I, Baier U, Khodjaniyazov E. Challenges and solutions for biogas production from agriculture waste in the Aral Sea basin. PRO. 2021;**9**(2):199. DOI: 10.3390/pr9020199

[20] Akhtar N, Iftikhar S, Shah H, Akhter N, Khan M, Tarar A, et al. Economic impact of biogas plants on the rural households: A case study of district Sargodha, Punjab-Pakistan. Transylvanian Review. 2017;**XXV**(21):5309-5313

[21] Regattieri A, Bortolini M, Ferrari E, Gamberi M, Piana F. Biogas micro-production from human organic waste—A research proposal. Sustainability. 2018;**10**(2):330. DOI: 10.3390/su10020330

[22] Anonymous. Provisional population totals. Paper-1, Census of India [Internet]. 2011. Available from: http://www.nird.org.in [Accessed: June 11, 2019]

[23] Ravindranath D, Rao S S N. Bio-energy in India: Barriers and policy options. [Internet]. 2015. Available from: https://ledsgp.org/app/uploads/2015/07/Bioenergy-in-India.pdf [Accessed: June 11, 2019]

[24] Anonymous. Biogas programme. No. 126/1/2023-Biogas, Ministry of New and Renewable Energy [Internet]. 2023. Available from: http://www.mnre.gov.in/bio-energy/schemes [Accessed: July 31, 2023]

[25] Maniatis K. Progress in biomass gasification: An overview. In: Bridgwater AV, editor. Progress in Thermochemical Biomass Conversion. Oxford, UK: Blackwell Science Ltd; 2001. pp. 1-31. DOI: 10.1002/9780470694954.ch1

[26] Painuly JP. Barriers to renewable energy penetration; A framework for analysis. Renewable Energy. 2001;**24**(1):73-89. DOI: 10.1016/S0960-1481(00)00186-5

[27] Soni R, Singh I. Producing electricity from biomass gasification. Progressive Farming. 2014;**50**(1):10-15

[28] Rao DM, Rajesh RM, Badhe M. Biomass gasification status in India. IEA Bioenergy. 2021;**33**:10-11

[29] Paustian K, Cole CV, Sauerbeck D, Sampson N. CO_2 mitigation by agriculture: An overview. Climatic Change. 1998;**40**(1):135-162. DOI: 10.1023/A:1005347017157

[30] Podkówka Z, Čermák B, Podkówka W, Broucek J. Greenhouse gas emissions from cattle. Ekologia (Bratislava). 2015;**34**(1):82-88. DOI: 10.1515/eko-2015-0009

Chapter 6

Bacteriophage as Biotechnological Tools to Improve the Effectiveness of Anaerobic Digestion Process

Şuheda Reisoglu and Sevcan Aydin

Abstract

Wastewater treatment plants (WWTPs) serve as habitats for diverse and densely populated bacterial communities, fostering intricate microbial interactions. Conventional treatment methods employed often fail to completely eliminate pathogens. Consequently, inadequate chemical treatments lead to the eventual release of waterborne bacterial pathogens into the environment through effluent water. Anaerobic digestion represents a biological treatment approach for organic waste and wastewater, providing cost-reduction benefits and enabling energy generation through biogas production from organic waste. However, the role of viruses-host interactions in anaerobic digestion and their effects on biological wastewater treatment (WWT) has been lacking and requires further research and attention. Bacteriophages (phages), viruses that target specific bacteria, are abundant within WWTPs and engage in diverse interactions with their host organisms. Also, there are reports indicating the presence of archaeal viruses capable of impacting crucial methanogenic organisms in anaerobic digestion, alongside phages. Despite their apparent lack of discernible metabolic functions, viral community have significant potential to influence WWT by shaping the structure of microbial communities, thereby impacting the efficiency of the processes. This chapter aims to explore the influence of reported viral communities, especially phages on shaping microbial communities; elucidate the dynamics and limitations of phage-host relationships; and evaluate their potential as biological tools for enhancing the anaerobic digestion process in WWT.

Keywords: anaerobic digestion, bacteriophage, biofilm removal, community dynamics, wastewater treatment

1. Introduction

Engineered biological systems comprise the basis of high-potential environmental processes such as wastewater treatment (WWT) and the production of bioenergy carriers. WWT is an application in which modified microbiological processes take a role in carbon and nutrient removal biologically, also termed biological wastewater treatment (BWT) systems such as anaerobic digesters. In this regard, anaerobic digestion (AD) is a useful and crucial process in which microbial communities comprise bacteria and archaea, which can degrade organic matter in anoxic conditions. AD is significantly employed in industrial processes besides occurring naturally in various environments including aquatic sediments and animal gut. It is a valuable process that plays an important role in reducing fossil fuel dependency and producing methane by transforming waste disposal into a useful process [1]. The anaerobic digestion process encompasses four primary stages: hydrolysis, acidogenesis, acetogenesis, and methanogenesis. Acetogenic bacteria play a crucial role in this process by converting simple substrates, including acetate, H_2, CO_2, and various fermentation by-products such as propionate, butyrate, and alcohols. The preservation of a specific archaeal group, known as methanogens, is imperative to ensure the successful and stable operation of the AD process. This group is responsible for catalyzing the terminal and most delicate step of the anaerobic process, which is methanogenesis. Methanogens are generally categorized into two main groups based on their substrate conversion capabilities: acetotrophic and hydrogenotrophic methanogens, and of particular importance, acetotrophic methanogens significantly contribute to methane production, with approximately 70% of the generated methane originating from acetate conversion [2].

Structural changes in bacterial communities frequently cause variations in the effectiveness of biological wastewater treatment (BWT) systems. Therefore, the stabilization of microbial composition is very important for BWT. Although BWTs primarily use bacterial communities to decompose contaminants, the AD composition in the treatment is quite diverse, extending beyond bacteria. Although a diverse array of organisms contributes to the intricate dynamics of the microbial community in this process, there is an increasing recognition of the role that viruses, particularly bacteriophages, play in controlling bacterial populations. Moreover, the viral concentration is significantly higher in samples from WWTPs compared to in other aquatic environments, and the strong correlation detected between bacteriophages and bacteria in BWTs adds to this interest. This interest has moved toward the role of bacteriophages in assessing process efficiency and waste quality [3].

Bacteriophages, or simply phages, are viruses that target and infect specific bacteria and, like other viruses, are mostly obligate parasites, so they do not have an internal metabolism and need the metabolic mechanism of their hosts to maintain their life cycle [4]. It has been revealed that phage activity in various aquatic and terrestrial ecosystems, especially in the oceans, is a driving force in shaping microbial communities and in biodiversity through interspecies gene transfer [5]. Concentrations of phage are expressed to be approximately 108–109 particles per milliliter in BWT harboring wastewater from a wide variety of different sources, and this means a higher concentration than other ecosystems studied so far (**Figure 1**) [6]. In BWT, there is a significant increase in interest in phages in terms of their impact on the bacterial consortium and consequently process efficiency and effluent quality. It was revealed that there is a relationship between the viral community structure and methane production in the anaerobic digesters of WWTs, and it was found that the viral shunt was effective in methane production [7]. Even though anaerobic digesters have the potential to study the relationship of phages to the microbial community and the effects of these relationships on the system, as they are systems in which a large number and variety of phage and prokaryotic host interactions are associated with easy access and nutrient-rich environment, the effects of phages on the microbial community are still not fully understood. This is since classical methods used in phage studies were generally culture-dependent, and because of bacteria that are still unculturable or difficult to culture in BWT, it is difficult or impossible to detect phages

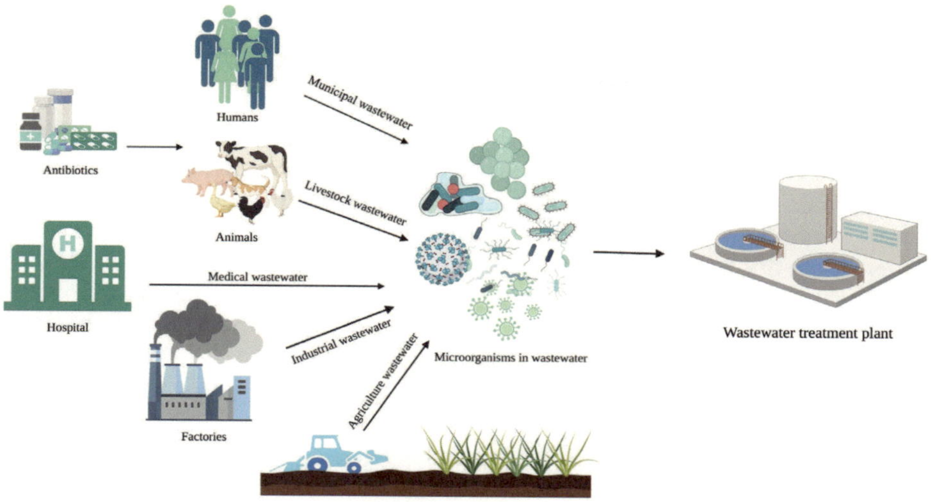

Figure 1.
Since wastewater treatment systems receive wastewater from different sources, various bacterial/viral/fungal microorganisms that come with these sources enter the treatment systems along with the wastewater.

specific to these bacteria by culture-dependent methods [8]. With the development of culture-independent methods, especially shotgun metagenomic with bioinformatic tools, novel phages are discovered in BWT systems and detailed information on phage-host dynamics is obtained [9]. The aim of this chapter is therefore to explain the role of phages in AD process by examining the effects of phages on anaerobic microbial community structure in the light of this new information.

2. Anaerobic microbial community dynamics and interaction

2.1 Phage-host dynamics in anaerobic digestion

The relationship of the phage with its host is based on the host density and infection frequency. Thus, these two parameters may be the driver of the evolution of phage-host dynamics in BWT. A study examining sequencing data from four anaerobic digesters in full-scale WWTPs revealed monthly fluctuations in phage and prokaryotic populations over a year, demonstrating significant correlations at both α- and β-diversity levels and supporting the notion that cell lysis operates in a density-dependent manner [7]. As to the other factor, high infection rates of virulent phages can result in excessive prokaryotic mortality, although the impact may vary due to differences in phage titers and infection cycles. In the context of BWT, an "arms race" ensues between virulent phages and their hosts, marked by increasing phage infectivity and host resistance over time. Prokaryotes employ various strategies to resist phage invasion, including protein-based defense mechanisms and emerging chemical antiphage defenses. Some of these mechanisms have been observed in WWTPs, demonstrating adaptation to local phage predation pressures. The "arms race" dynamics, characterized by ongoing host-phage interactions, incur costs for both bacterial resistance and phage infectivity resistance. As a result, these dynamics may eventually transition to fluctuating selection dynamics, where phage and host genotype frequencies oscillate due to negative frequency-dependent selection. This may provide an advantage to rare bacterial resistance alleles *via* phage evolution to infect common bacterial genotypes [10].

In addition to the host density and infection, the type of infection also has the capacity to seriously affect population dynamics and the functions of organisms. Phages generally affect their host in two ways: lytic and lysogenic. The lytic cycle results in the death (lysis) of the host bacteria following infection and the subsequent release of new phage virions, resulting in a drastic decrease in bacterial population density in this type of infection. In lysogeny, the phage cycle is continued by integrating phage genetic material within the genetic material of the host (prophage formation) without the lysis of the host cell. This means that the bacterial density in the medium is not changed by the phage, since there is no lysis of the host cells (**Figure 2**).

In a study showing that phages affect microbial community structure and performance of the process in BWT systems, *Microlunatus phosphovorus* density in an activated sludge reactor diminished with the addition of phages [11]. In another study conducted in an MBR system receiving industrial wastewater, phage abundance showed an inverse proportion to both the bacterial hosts and related bacteria [3]. Yang et al. [12] detected a simultaneous decrease in phage concentration and diversity with the increase in bacterial concentration in activated sludge in the municipal wastewater system. Brown et al. [13] stated that while there was a substantial relationship between increasing virus abundance and decrease in bacterial abundance in the nitrification process, the abundance of viral particles was notably affected by pH and magnesium ion exchange, which are effective in the attachment of phages to the host cell.

While lytic phage infection is characterized by a marked decrease in the number of host bacteria, the prophage incorporated into the bacterial genome rather than killing

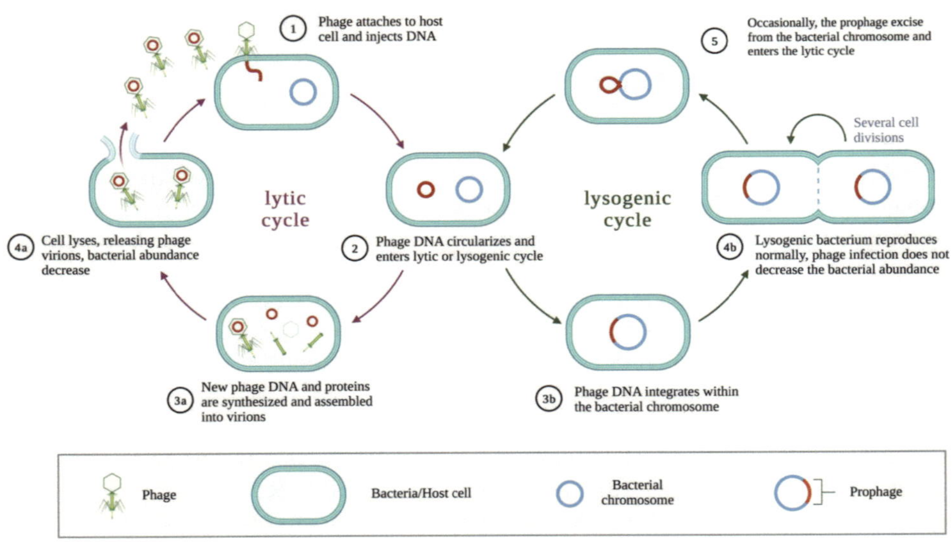

Figure 2.
The lytic and lysogenic life cycle of bacteriophages. When bacteria are infected by lytic phages, the bacterial cells undergo lysis, and the abundance of bacteria decreases while lysogenic infection does not decrease the infected bacterial abundance.

the bacterium may benefit the bacterium and promote its growth. Incorporation of the prophage into the host genome can protect the bacterial host by developing resistance to lytic phage infection, increase pathogenicity by promoting host growth and spread, and may provide the host with an advantage over the competition with other strains [14]. Prophages can support the resistance and bacterial stability in the microbial population, rather than cause the infected population to decline [15]. On the other hand, Heyer et al. [16] revealed that lytic phage infection can reduce the biogas production amount in the anaerobic digester while lysing the host cell, increasing nutrient cycling and promoting the growth of auxotrophic bacteria.

In addition to the other microbial communities in wastewater treatment, archaeal community dynamics have crucial importance for improving the anaerobic treatment and waste reclamation *via* biogas. In a study conducted by Aydin et al. [9], the density of the archaeal community and structure were evaluated *via* Illumina Next-Generation Sequencing. Before the implementation of phage cocktails in the reactor, the most abundant archaeal genera were found to be *Methanothermobacter* and *Methanosaeta*; however, microbial dynamics and archaeal community underwent alterations with the addition of the phage cocktail. While the dominant genus of the archaeal composition in phage-added reactor was *Methanoculleus* with a 43% ratio, both *Methanosaeta* and *Thermoplasmatales* followed with 22% relative abundance. Even though abundances of methanogens show alterations in the AnMBRs, the main pathway in the production of biogas continued as hydrogenotrophic methanogenesis.

2.2 Archaeal viruses in anaerobic digestion

Viruses exhibit a ubiquitous presence within the biosphere, exerting significant influence upon the hosts they infect. A widely accepted concept is that viruses can target members of virtually every microbial taxon, with an extensive catalog of bacterial viruses, namely, phages, already well-documented. In addition to phages, there are viruses known to infect archaeal cells, termed archaeal viruses or archeoviruses. The ramifications of viral interactions for the structural composition and population dynamics of archaeal communities within anaerobic digesters have only recently emerged as a focal point of scientific inquiry, representing a field that remains largely uncharted [17]. Investigations into archaeal viruses, in contrast to phages, reveal relatively limited information; nevertheless, various archaeal viruses have been reported to infect microorganisms associated with anaerobic digesters. In a study conducted by Wolf et al., a lytic archaeal virus named "Drs3" was detailed alongside its host, *Methanobacterium formicicum* strain Khl10. This hydrogenotrophic methanogenic archaeon and its corresponding virus were both obtained from the anaerobic digester of an experimental biogas plant [18].

Based on the current scientific literature, the existence of mycoviruses, viruses that target fungal cells, within anaerobic digesters has not been ascertained. However, the absence of reports does not preclude the possibility of mycoviruses inhabiting these environments. Indeed, it is plausible that mycoviruses remain largely unexplored and yet to be discovered, representing a promising frontier in the quest to unravel the intricate dynamics governing microbial communities within anaerobic digestion systems. Drawing from the widely accepted concept that viruses have the capacity to infect members of virtually every microbial taxon, the potential existence of mycoviruses underscores the imperative need for advanced research endeavors to extend the comprehension of this intricate ecosystem [17]. Utilizing isolation-independent approaches, like metagenomics and genome-based searches, has the potential to augment the catalog of viruses linked to methanogenic archaea in anaerobic digesters. Through the detection of viral communities and their interplay with methanogens, there is the prospect of improvement of the stability of anaerobic digesters. By unraveling the presence and dynamics of individual viral communities in conjunction with methanogens, there is a potential avenue to bolster the stability of anaerobic digesters [19].

3. Pathogen biocontrol

WWTPs receive wastewater from different sources such as municipal, industrial, and livestock wastewater in addition to hospitals, which include highly opportunistic pathogenic bacteria. Therefore, the influent water of BWT generally involves pathogens including opportunistic *Escherichia coli* strains, *Salmonella spp*., *Staphylococcus aureus*, and *Acinetobacter baumannii*. Despite using different biological or chemical water treatment methods, WWTPs are not fully capable of eliminating pathogens, so pathogenic species can often be found in the effluent, threatening environmental and public health through release in the aquatic environments [20]. For instance, the opportunistic pathogen *A. baumanni* was detected in each stage of WWTP and eventually released from the treatment plant *via* effluent water to the aquatic media without lysis [21]. Also, it has been indicated by Oliveira et al. [22] that the conventional treatment process for carbapenem-resistant pathogen removal was not adequate with the discovery of this pathogen in effluent water samples. The effluent water, which is reused as drinking water or for recreational purposes, may contain microbial contamination due to the lack of adequate and effective treatment. If wastewater discharges contain fecally transmitted pathogens originating from humans and livestock, diseases can spread even in developed countries with the reuse of this treated water [23]. Considering the high lethal levels of pathogenic bacteria found in discharged effluent samples without any degradation and their dissemination into the environment through aquatic resources, it is obvious that pathogen removal must be ensured as much as possible with the most efficient approaches in the BWT process.

In recent years, the phage application for pathogen biocontrol has been gaining substantial attention instead of the current physicochemical methods since phages can be active until the last target bacterial cell is eliminated [24]. The use of antibiotics in the treatment of pathogens from wastewater may not give the desired result, since most pathogens have developed resistance to antibiotics. Phages have several properties that make them appealing as therapeutic or biocontrol agents. While phages have an antibacterial effect in removing antibiotic-resistant pathogens, they are equally influential for antibiotic-resistant bacteria. Phages, which target and lysis only bacteria specific to them through special receptors, can increase in number depending on the density of pathogens and can easily adapt to the environment [25]. Self-limiting phages hardly affect the native flora. Being naturally found in any environment, they are generally easily discovered and are easy to isolate, especially with culture-independent methods [26].

Recently, pathogen-specific phage isolation has been made and some of these isolated phages have been found to be successful agents of pathogen biocontrol in BWT. In a study conducted by Dhevagi and Anusuya [27], the application of *E. coli* and *Salmonella* phages resulted in a significant reduction in the abundance of these pathogens in sewage sludge. In addition to single phage application, the application of a phage cocktail, in which more than one species-specific phage is combined, is a popular phage application method since it expands the host range and eliminates the development of phage resistance and the pathogen recovery after phage implementation [28]. In this regard, the efficiency of a single phage, a combination of a two-phage, and a cocktail of three phages on the removal of *Salmonella* in wastewater were tested. The phage cocktail provided the most efficient removal of pathogens compared to the other two [29]. In addition, phage cocktail consisting of polyvalent phages with a wider host range can exhibit more successful results. For example, in a study conducted in activated sludge systems, it was determined that the cocktail consisting of polyvalent phages was more effective than the cocktail consisting of phages with narrow host ranges in reducing the antibiotic-resistant *E. coli* strain [30]. It is worth mentioning that a careful selection of phages to be used in pathogen biocontrol is crucial to targeting and eliminating bacterial populations. Because lysogenic phages do not always degrade the bacterial cell and may undesirably transfer genetic information *via* horizontal gene transfer, therefore lysogenic phages may not be suitable for pathogen biocontrol. The phages to be used for this purpose must have a lysis cycle [31].

4. Biofilm disaggregation

Biofilm is a complex matrix structure that bacteria produce in their environments and infections in order to survive. Bacteria forming the biofilm are embedded in the extracellular polymeric substance (EPS) matrix. These EPSs are typically polymeric substances consisting of polysaccharides, proteinaceous substances, glycopeptides, lipids, and lipopolysaccharides. The biofilm, which develops as bacteria adhere to the surface and form colonies, is usually a complex structure consisting of the combination of more than one bacterial species and acts as a protective shield for bacteria against antibacterials and other bacteria. Through this protective structure, the stability of the bacteria increases while at the same time reducing the effectiveness of traditional antibacterial agents. For membrane bioreactors, which is a novel approach in wastewater treatment, biofilm structures form challenges, because the biofilm layer significantly affects the normal operation of the device by continuously reducing the net wastewater flow rate passing through the membrane surface. In recent years, phage-based applications as a solution to the biofilm-based blockage and foaming problem have attracted intense interest from researchers due to the unresponsiveness of resistant bacteria to existing antibiotics [32]. In this context, there are studies in which phages are used in the disaggregation of biofilms, which are formed by pathogenic bacteria in wastewater treatment and limit the progress and effect of the process. By using the lytic phage isolated from the wastewater treatment system to eradicate the biofilm formed by the *Delftia tsuruhatensis* pathogen on the membrane filter, the membrane flow was improved by 70% and the success of the isolated phage as a biocontrol agent was demonstrated [33]. In the context of an urban wastewater treatment facility, approximately 48 bacteriophages specific to *Proteus mirabilis* were isolated, thereby inhibiting the development of biofilm formation caused by this strain [34]. Ayyaru et al. [35] stated that the *E. coli* phage they used to clean the nanocomposite membrane contaminated by antibiotic-resistant pathogens solved the contamination problem and increased membrane flow. Another illustration of biofilm disaggregation was provided in a study conducted by [36], which demonstrated the inhibitory activity of bacteriophage MA-1 against various strains of *Pseudomonas aeruginosa*.

The interaction between the bacteria in the biofilm and the related phages of these bacteria is seen as an important tool in the fight against biofilms formed by bacteria. While bacteria use EPSs to protect them from attack by phages, it becomes easier for phages that can express polysaccharide depolymerase enzymes to find the appropriate host receptor (**Figure 3**) [38]. In this context, genetically engineered phages can provide a high rate of removal in biofilm degradation. It has been shown that the modified T7 phage shows very high success in biofilm removal by overexpressing DspB, a polysaccharide depolymerase enzyme [39]. As in eliminating the formation of pathogenic bacteria in complex structures, the application of a phage cocktail containing more than one type of phage may be more beneficial than the application

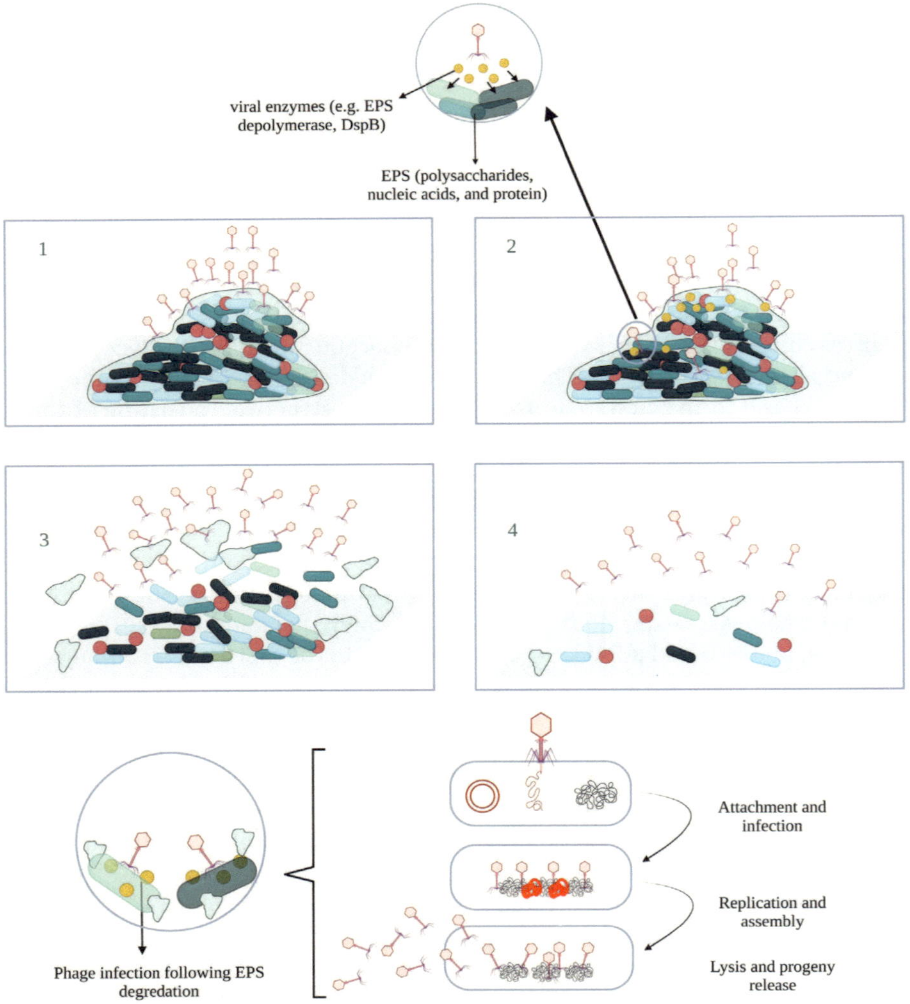

Figure 3.
Bacterial biofilm disaggregation through phage-mediated disruption. As depicted in the figure, the sequential steps involved in biofilm biocontrol using bacteriophages comprise (1) phage application, (2) initiations of EPS disruption through enzymes such as dispersion B (DspB) or EPS depolymerizes, (3) biofilm disaggregation through the disintegration of EPS and exposure of the phage infection, and (4) completion biofilm disaggregation and cell lysis [37].

of a single phage in biofilm removal. Because biofilm structures are usually complex structures formed by more than one type of bacteria, treatment with a wide host range is essential to interfere with this structure. Wide-host-range polyvalent phages and phage cocktails are an important strategy in biofilm disaggregation, offering high efficiency in recent years [40].

5. Reconstruction of microbial community structure

The microbial community's composition is influenced by various factors, including both biotic and abiotic elements. However, our understanding of biotic control remains limited due to the lack of suitable tools for studying biological interactions. Recent studies have highlighted the significant presence of phages, which are recognized as the most abundant biological entities on Earth [41]. Consequently, there has been growing interest in investigating the impact of bacteriophage predation on microbial communities. For instance, in the oceans, phages are responsible for lysing more than 25% of microbial cells. Similarly, in natural freshwater environments, phage lysis accounts for up to 71% of microbial mortality, exerting a profound influence on microbial food webs and the cycling of aquatic carbon [42].

In the context of WWTPs, anaerobic BWT represents a highly dynamic process wherein the symbiotic relationship between bacteria and archaea leads to the production of methane, a significant bioenergy source. The utilization of phages and their beneficial application in the augmentation of anaerobic bioreactor systems emerges as a promising strategy for enhancing performance, mitigating membrane biofouling, and influencing community shape, as illustrated in **Figure 3**. By selectively influencing specific bacteria within the community using phages, it is plausible to enhance methane production by modulating the proportion of archaea. The relationship between bacteria and archaea in anaerobic environments, such as anaerobic bioreactors, is vital for methane production. Bacteria break down complex organic compounds into simpler forms, which are then utilized by archaea to produce methane. Through the introduction of bacteriophages, which target and control the growth of specific bacterial strains, it is possible to manipulate the composition and dynamics of the microbial community. By reducing the abundance of certain bacteria through phage-mediated mechanisms, it is conceivable to shift the balance within the community toward favoring specific archaeal populations that are more efficient in methane production [43]. This targeted modulation of the microbial community can optimize metabolic processes and potentially enhance methane production rates.

The outcome of phage-host interactions and their influence on the coexistence dynamics of bacterial hosts within microbial communities depend on the nature of these interactions as well as external conditions. The nature of phage-host interactions refers to factors such as the specificity of phages toward certain bacterial strains, the efficiency of phage infection and replication within hosts, and the ability of hosts to develop resistance mechanisms against phages. These factors determine whether phages act as facilitators of coexistence or promoters of competitive exclusion. External conditions, including environmental factors and resource availability, also play a significant role. Furthermore, the presence of other species within the community can interact with phage-host dynamics, influencing the outcome. Interactions between phages, bacterial hosts, and other community members can create complex ecological networks that shape coexistence patterns [44]. Overall, the coexistence or competitive exclusion effects mediated by phages within microbial communities are contingent upon the specific phage-host interactions and the prevailing external conditions. Understanding these factors is essential for comprehending the dynamics and stability of microbial communities. Every instance of phage infection introduces novel genetic information to the specific bacteria it targets, establishing a close and interdependent relationship between bacteriophages and their host bacteria. This dynamic connection between phages and hosts plays a pivotal role in shaping the evolution and ecology of microbial communities [45].

The interaction between phages and bacteria can reshape bacterial communities through various mechanisms. It can alter the dynamics of bacterial competition, drive bacterial diversity, or facilitate horizontal gene transfer between different species within the microbial community [46]. Scientists have proposed several approaches to explain this coevolutionary relationship. One of these approaches is the Red Queen hypothesis. This hypothesis, originally proposed by Van Valen [47], suggests that the successful adaptation of one organism can reduce the adaptability of other species that inhabit the same environment and interact with it. In the context of phages and bacteria, this hypothesis involves a coadaptation process referred to as an "arms race," wherein these entities constantly engage in defensive strategies against each other [48]. However, when the level of resistance between phages and their host bacteria reaches a high threshold, the ongoing arms race between them may deteriorate, ultimately leading to the emergence of fluctuating selection dynamics. This phenomenon can create an opportunity for rare bacterial resistance alleles to persist and exploit the evolutionary dynamics of phages, enabling them to infect more prevalent bacterial genotypes [49].

The second hypothesis, Kill-the-Winner, proposes that the dynamics between phages and bacterial hosts are driven by negative frequency-dependent selection within the microbial population. This hypothesis suggests that phages selectively target and infect the most abundant bacterial strains, often referred to as the "winners" in the population. By reducing the population size of these dominant strains, phages alleviate competition and create opportunities for less abundant strains, or "losers," to proliferate. The negative frequency-dependent selection arises from the fact that the efficacy of phages in infecting bacterial hosts depends on the relative abundance of the target strains. The more abundant a particular bacterial strain becomes, the higher its susceptibility to phage infection is due to the increased encounter rate between the phages and their hosts. As a result, the growth of dominant strains is suppressed, allowing less abundant strains to catch up and contribute to the overall diversity of the microbial community. This mechanism of negative frequency-dependent selection promotes the coexistence of multiple bacterial strains by preventing any one strain from achieving long-term dominance. Instead, it maintains a dynamic equilibrium where the relative abundance of bacterial strains fluctuates over time [50]. The Kill-the-Winner hypothesis highlights the role of phages in shaping the structure and diversity of microbial communities by balancing the competitive interactions among bacterial strains. It emphasizes the importance of considering the ecological and evolutionary dynamics between phages and their bacterial hosts in understanding the stability and functioning of microbial ecosystems. In resource-limited environments, phages may act as "Kill-the-Winner" agents, targeting and reducing the population of dominant bacterial strains, which allows less abundant strains to thrive and coexist. In contrast, under resource-rich conditions, phages may selectively infect and eliminate specific strains, leading to competitive exclusion. This recurring rise and fall of specific microbial populations can contribute to bacterial diversity by regulating the competitiveness of different species and facilitating the coexistence of diverse organisms [51].

The third explanation for phage-bacteria interactions is the concept of polyvalency. It is commonly observed that phages exhibit a narrow host range, meaning they can only infect specific bacterial species or strains. However, there are exceptions to this pattern, as certain phages known as polyvalent phages possess a broad host range, enabling them to infect multiple bacterial species. Polyvalent phages have the ability to recognize and bind to a wider range of host receptors, allowing them to infect diverse bacterial hosts. This broader host range is advantageous for phages in environments where multiple bacterial species coexist, as it increases their potential targets for infection [52]. The existence of polyvalent phages has been documented in various studies, and their broad host range has significant implications for microbial communities. By infecting multiple bacterial species, polyvalent phages can influence community dynamics, including competition and coexistence patterns, as well as the overall stability and diversity of the microbial community. Understanding the prevalence and impact of polyvalent phages adds a layer of complexity to the interplay between phages and bacterial hosts within microbial communities [53].

The impact of phage-driven changes in microbiota on BWT performance has been observed in WWTPs. Liu et al. [54] addressed the knowledge gap regarding phage population dynamics during sludge bulking. They noted a substantial reduction in the abundance of nitrifying bacteria during sludge bulking. Moreover, viral contigs linked to nitrifiers were more frequently identified and found in greater abundance in viromes from bulking sludge samples. These findings suggest that phages may contribute to the decline of autotrophic nitrifiers under bulking conditions. Additionally, through the utilization of advanced sequential methodologies, a significant revelation emerged regarding the phages isolated from activated sludge. It was discerned that a subset of these phages demonstrated a notable interspecies infection capability, signifying their capacity to infect bacteria across diverse species boundaries [30]. In the study of Aydin et al. [9], the effects of a phage cocktail on the treatment of pharmaceutical wastewater using an anaerobic bioreactor were investigated. The implementation of the phage cocktail resulted in significant changes in the bacterial and archaeal community in both the biofilm and sludge. Notably, there was a transition from *Methanothermobacter* to *Methanoculleus* as the dominant archaeal community, accompanied by a syntrophic interaction between the bacterial genera *Macellibacteroidetes-Desulfovibrio* and the archaeal genus *Methanoculleus*. These findings suggest that harnessing bacteriophages could be a promising strategy for regulating bacteria within anaerobic microbial communities and restoring the balance between bacterial and archaeal populations in a rapid manner. However, it is important to note that there is a limited number of experimental studies exploring the effects of different bacteriophage species and combinations on microbial community structure and activity within anaerobic bioreactors. Hence, conducting comprehensive studies over extended operational periods is strongly recommended to assess the influence of different phage species and combinations on the performance of anaerobic reactors and the dynamics of microbial communities. In conclusion, this study highlights the potential of phage-based approaches in enhancing the treatment of pharmaceutical wastewater in anaerobic bioreactors. Further research is needed to explore a broader range of bacteriophage species and combinations, as well as to assess their long-term effects on reactor performance and microbial community dynamics. Such studies will contribute to advancing our understanding and application of phage-mediated strategies in anaerobic wastewater treatment.

6. Limitations and perspectives

Typically, phages are thought to target specific bacterial species or strains. However, polyvalent phages with a broad host range have been frequently found in wastewater treatment systems. Additionally, multiple phages may infect hosts that are more abundant in the environment. Consequently, phage-host interactions in wastewater treatment ecosystems often form intricate networks. Yet it remains challenging to identify phage-host relationships under *in situ* conditions [53]. Traditional host-range assessments rely on available hosts in laboratory settings. However, in complex ecosystems like wastewater treatment systems, the diversity and number of potential hosts far exceed those found in labs, and phages may also compete for hosts. While modern computational methods based on sequencing can predict host interactions, these methods typically establish one-to-one relationships [55]. Therefore, they may not fully capture the complex phage-host infection dynamics in wastewater treatment systems, necessitating the development of innovative approaches to assess host ranges under *in situ* conditions.

Furthermore, unlike virulent phages, temperate phages have the ability to make a pivotal decision shortly after infecting a host cell: they can opt for lytic growth, causing the host cell to burst, or they can enter the lysogenic cycle, integrating their genetic material into the host cell as a prophage. Given that free phages are consistently flushed out with the effluent, prophages might have a higher likelihood of persisting in BWT systems. Prophages can play a direct role in host cell survival in unfavorable conditions by suppressing unnecessary metabolic activities. However, they also carry the potential to act as genetic bombs that can lead to host cell lysis, or even the demise of the entire host population, under specific environmental circumstances [56]. Extensive prophage induction followed by the sudden lysis of a substantial portion of the microbial community can contribute to issues like foaming, bulking, or reduced process efficiency in BWT systems [57]. Nevertheless, temperate phages in BWT environments have not been comprehensively investigated. The factors that influence the choice between lytic and lysogenic infections remain poorly understood, and the triggers for prophage induction are yet to be explored.

7. Conclusion

Phages possess the capacity to influence the diversity and arrangement of microbial communities within anaerobic digesters of wastewater treatment plants (WWTPs). The occurrence of frequent fluctuations in both phages and prokaryotes has the potential to impact the stability of the microbial composition and evolutionary processes. However, it is important to note that phages are not the sole viral entities present in anaerobic digesters. There have been reports of archaeal viruses with the potential to influence the methanogenic organisms critical to the anaerobic digestion process. Considering the possibility that each species may be susceptible to one or more viruses, it is reasonable to assume that the current knowledge of viruses affecting methanogens is merely the tip of the iceberg. Therefore, it is imperative to conduct further investigations into the genetic diversity of viruses targeting methanogens.

Of equal importance is a comprehensive understanding of the dynamics governing virus-host interactions throughout the AD process. Furthermore, the deliberate implementation of host-based treatments, tailored to regulate the prevalence of distinct microbial groups, holds promise for effectively addressing challenges like bulking and foaming issues in the process. These strategic interventions can guide the system toward its intended operational objectives. Enhancing the understanding of how these viruses impact the microbial community of AD and its dynamics will play a pivotal role in evaluating the efficiency and stability of the entire biogas production process.

Author details

Şuheda Reisoglu and Sevcan Aydin*
Division of Biotechnology, Biology Department, Faculty of Science, Istanbul University, Istanbul, Turkey

*Address all correspondence to: sevcan.aydin@istanbul.edu.tr

References

[1] Dutta S, He M, Xiong X, Tsang DC. Sustainable management and recycling of food waste anaerobic digestate: A review. Bioresource Technology. 2021;341-348. DOI: 10.1016/j.biortech.2021.125915

[2] Aydin S, Ince B, Ince O. Application of real-time PCR to determination of combined effect of antibiotics on Bacteria, Methanogenic Archaea, Archaea in anaerobic sequencing batch reactors. Water Research. 2015;**76**:88-98. DOI: 10.1016/j.watres.2015.02.043

[3] Shapiro OH, Kushmaro A, Brenner A. Bacteriophage predation regulates microbial abundance and diversity in a full-scale bioreactor treating industrial wastewater. The ISME Journal. 2010;**4**:327-336. DOI: 10.1038/ismej.2009.118

[4] Elois MA, da Silva R, Pilati GVT, Rodríguez-Lázaro D, Fongaro G. Bacteriophages as biotechnological tools. Viruses. 2023;**15**(2):349. DOI: 10.3390/v15020349

[5] Suttle CA. Marine viruses—Major players in the global ecosystem. Nature Reviews Microbiology. 2007;**5**(10):801-812. DOI: 10.1038/nrmicro1750

[6] Du B, Yang Q, Li X, Yuan W, Chen Y, Wang R. Impacts of long-term exposure to tetracycline and sulfamethoxazole on the sludge granules in an anoxic-aerobic wastewater treatment system. Science of the Total Environment. 2019;**684**:67-77. DOI: 10.1016/J.SCITOTENV.2019.05.313

[7] Zhang J, Gao Q, Zhang Q, Wang T, Yue H, Wu L, et al. Bacteriophage–prokaryote dynamics and interaction within anaerobic digestion processes across time and space. Microbiome. 2017;**5**(1):57. DOI: 10.1186/s40168-017-0272-8

[8] Edwards RA, Mcnair K, Faust K, Raes J, Dutilh BE. Computational approaches to predict bacteriophage-host relationships. FEMS Microbiology Reviews. 2016;**048**:258-272. DOI: 10.1093/femsre/fuv048

[9] Aydin S, Can K, Çalışkan M, Balcazar JL. Bacteriophage cocktail as a promising bio-enhancer for methanogenic activities in anaerobic membrane bioreactors. Science of the Total Environment. 2022;**832**:154716. DOI: 10.1016/J.SCITOTENV.2022.154716

[10] Azam AH, Tanji Y. Bacteriophage-host arm race: An update on the mechanism of phage resistance in bacteria and revenge of the phage with the perspective for phage therapy. Applied Microbiology and Biotechnology. 2019;**103**(5):2121-2131. DOI: 10.1007/s00253-019-09629-x

[11] Lee SH, Otawa K, Onuki M, Stoh H, Mino T. Population dynamics of phage-host system of microlunatus phosphovorus indigenous in activated sludge. Journal of Microbiology and Biotechnology. 2007;**17**(10):1704-1707

[12] Yang Q, Zhao H, Du B. Bacteria and bacteriophage communities in bulking and non-bulking activated sludge in full-scale municipal wastewater treatment systems. Biochemical Engineering Journal. 2017;**119**:101-111. DOI: 10.1016/J.BEJ.2016.12.017

[13] Brown MR, Baptista JC, Lunn M, Swan DL, Smith SJ, Davenport RJ, et al. Coupled virus-bacteria interactions and ecosystem function in an engineered

microbial system. Water Research. 2019;**152**:264-273. DOI: 10.1016/j.watres.2019.01.003

[14] Fortier L-C, Sekulovic O. Importance of prophages to evolution and virulence of bacterial pathogens. Virulence. 2013;**5**:354-365. DOI: 10.4161/viru.24498

[15] Kim MS, Park EJ, Roh SW, Bae JW. Diversity and abundance of single-stranded DNA viruses in human feces. Applied and Environmental Microbiology. 2011;**77**(22):8062-8070. DOI: 10.1128/AEM.06331-11

[16] Heyer R, Schallert K, Siewert C, Kohrs F, Greve J, Maus I, et al. Metaproteome analysis reveals that syntrophy, competition, and phage-host interaction shape microbial communities in biogas plants. Microbiome. 2019;**7**:1-17. DOI: 10.1186/s40168-019-0673-y

[17] Harirchi S, Wainaina S, Sar T, Ali Nojoumi S, Parchami M, Parchami M, et al. Microbiological insights into anaerobic digestion for biogas, hydrogen or volatile fatty acids (VFAs): A review. Bioengineered. 2022;**13**(3):6521-6557. DOI: 10.1080/21655979.2022.2035986

[18] Wolf S, Fischer MA, Kupczok A, Reetz J, Kern T, Schmitz RA, et al. Characterization of the lytic archaaeal virus Drs3 infecting *Methanobacterium formicicum*. 2019;**164**:667-674. DOI: 10.1007/s00705-018-04120-w

[19] Calusinska M, Marynowska M, Goux X, Lentzen E, Delfosse P. Analysis of ds DNA and RNA viromes in methanogenic digesters reveals novel viral genetic diversity. Environmental Microbiology. 2016;**18**(4):1162-1175. DOI: 10.1111/1462-2920.13127

[20] Kristensen JM, Nierychlo M, Albertsen M, Nielsen PH. Bacteria from the genus Arcobacter are abundant in effluent from wastewater treatment plants. Applied and environmental microbiology. 2020;**86**(9):e03044-19. DOI: 10.1128/AEM.03044-19

[21] Higgins PG, Hrenovic J, Seifert H, Dekic S. Characterization of Acinetobacter baumannii from water and sludge line of secondary wastewater treatment plant. Water Research. 2018;**140**:261-267. DOI: 10.1016/J.WATRES.2018.04.057

[22] Oliveira M, Leonardo IC, Nunes M, Silva AF, Barreto Crespo MT. Environmental and pathogenic carbapenem resistant bacteria isolated from a wastewater treatment plant harbour distinct antibiotic resistance mechanisms. Antibiotics. 2021;**10**(9):1118-1130. DOI: 10.3390/antibiotics10091118

[23] Salvador D, Caeiro MF, Serejo F, Nogueira P, Carneiro RN, Neto C. Monitoring waterborne pathogens in surface and drinking waters. Are water treatment plants (WTPs) simultaneously efficient in the elimination of enteric viruses and Fecal indicator bacteria (FIB)? Water. 2020. DOI: 10.3390/w12102824

[24] Hassan AY, Lin JT, Ricker N, Anany H. The age of phage: Friend or foe in the new Dawn of therapeutic and biocontrol applications? Pharmaceuticals. 2021. DOI: 10.3390/ph14030199

[25] Torres-Barceló C. The disparate effects of bacteriophages on antibiotic-resistant bacteria. Emerging Microbes & Infections. 2018;**7**(1):1-12. DOI: 10.1038/s41426-018-0169-z

[26] Jofre J, Muniesa M. Bacteriophage isolation and characterization: Phages of Escherichia coli. Horizontal Gene Transfer: Methods

and Protocols. 2020:61-79. DOI: 10.1007/978-1-4939-9877-7_4

[27] Dhevagi P, Anusuya S. Bacteriophage based pathogen reduction in sewage sludge nature. Environment and Pollution Technology an International Quarterly Scientific Journal. 2013;**12**:7-16. Available from: www.neptjournal.com

[28] Abedon ST, Danis-Wlodarczyk KM, Wozniak DJ. Phage cocktail development for bacteriophage therapy: Toward improving spectrum of activity breadth and depth. Pharmaceuticals. 2021;**14**(10):1019-1043. DOI: 10.3390/ph14101019

[29] Turki Y, Ouzari H, Mehri I, Ammar AB, Hassen A. Evaluation of a cocktail of three bacteriophages for the biocontrol of Salmonella of wastewater. Food Research International. 2012;**45**(2):1099-1105. DOI: 10.1016/j.foodres.2011.05.041

[30] Yu P, Mathieu J, Li M, Dai Z, Alvarez PJJ. Isolation of polyvalent bacteriophages by sequential multiple-host approaches. Applied and Environmental Microbiology. 2016;**82**(3):808-815. DOI: 10.1128/AEM.02382-15

[31] Schmelcher M, Loessner MJ. Application of bacteriophages for detection of foodborne pathogens. Bacteriophage. 2014;**4**(2):e28137. DOI: 10.4161/bact.28137

[32] Wu B, Wang R, Fane AG. The roles of bacteriophages in membrane-based water and wastewater treatment processes: A review. Water Research. 2017;**110**:120-132. DOI: 10.1016/J.WATRES.2016.12.004

[33] Bhattacharjee AS, Choi J, Motlagh AM, Mukherji ST, Goel R. Bacteriophage therapy for membrane biofouling in membrane bioreactors and antibiotic-resistant bacterial biofilms. Biotechnology and Bioengineering. 2015;**112**:1644-1654. DOI: 10.1002/bit.25574/abstract

[34] Maszewska A, Wójcik E, Ciurzyńska A, Wojtasik A, Piątkowska I, DastychJ,et al.Differentiationofpolyvalent bacteriophages specific to uropathogenic Proteus mirabilis strains based on the host range pattern and RFLP. Acta Biochimica Polonica. 2016;**63**(2):10.18388/abp.2015_1114

[35] Ayyaru S, Choi J, Ahn Y-H. Biofouling reduction in a MBR by the application of a lytic phage on a modified nanocomposite membrane. Environmental Science Water Research & Technology. 2018;**4**:1624. DOI: 10.1039/c8ew00316e

[36] Adnan M, Ali Shah MR, Jamal M, Jalil F, Andleeb S, Nawaz MA, et al. Isolation and characterization of bacteriophage to control multidrug-resistant *Pseudomonas aeruginosa* planktonic cells and biofilm. Biologicals. 2020;**63**:89-96. DOI: 10.1016/j.biologicals.2019.10.003

[37] Harada LK, Silva EC, Campos WF, Del Fiol FS, Vila M, Dąbrowska K, et al. Biotechnological applications of bacteriophages: State of the art. Microbiological Research. 2018;**212-213**:38-58. DOI: 10.1016/j.micres.2018.04.007

[38] Pires DP, Melo LDR, Vilas Boas D, Sillankorva S, Azeredo J. Phage therapy as an alternative or complementary strategy to prevent and control biofilm-related infections. Current Opinion in Microbiology. 2017;**39**:48-56. DOI: 10.1016/J.MIB.2017.09.004

[39] Lu TK, Collins JJ. Disperasing biofilms with engineered enzymatic

bacteriophage. 2007. Available from: www.pnas.orgcgidoi10.1073pnas.0704624104

[40] Mathieu J, Yu P, Zuo P, Da Silva MLB, Alvarez PJJ. Going viral: Emerging opportunities for phage-based bacterial control in water treatment and reuse. Accounts of Chemical Research. 2019;**52**(4):849-857. DOI: 10.1021/acs.accounts.8b00576

[41] Chevallereau A, Pons BJ, van Houte S, Westra ER. Interactions between bacterial and phage communities in natural environments. Nature Reviews Microbiology. 2022;**20**(1):49-62. DOI: 10.1038/s41579-021-00602-y

[42] Bouvier T, Del Giorgio PA. Key role of selective viral-induced mortality in determining marine bacterial community composition. Environmental Microbiology. 2006;**9**(2):287-297. DOI: 10.1111/j.1462-2920.2006.01137.x

[43] Reisoglu Ş, Aydin S. Bacteriophages as a promising approach for the biocontrol of antibiotic resistant pathogens and the reconstruction of microbial interaction networks in wastewater treatment systems: A review. In: Science of the Total Environment. Vol. 890. Elsevier B.V; 2023. DOI: 10.1016/j.scitotenv.2023.164291

[44] Tang X, Fan C, Zeng G, Zhong L, Li C, Ren X, et al. Phage-host interactions: The neglected part of biological wastewater treatment. Water Research. 2022;**226**:119183. DOI: 10.1016/j.watres.2022.119183

[45] Koskella B, Brockhurst MA. Bacteria–phage coevolution as a driver of ecological and evolutionary processes in microbial communities. FEMS Microbiology Reviews. 2014;**38**(5): 916-931. DOI: 10.1111/1574-6976.12072

[46] Touchon M, Bernheim A, Rocha EP. Genetic and life-history traits associated with the distribution of prophages in bacteria. The ISME Journal. 2016;**10**:2744-2754. DOI: 10.1038/ismej.2016.47

[47] Van Valen L. Molecular evolution as predicted by natural selection. Journal of molecular evolution. 1974;**3**:89-101

[48] Buckling A, Rainey PB. Antagonistic coevolution between a bacterium and a bacteriophage. Proceedings of the Royal Society B: Biological Sciences. 2002;**269**(1494):931-936. DOI: 10.1098/rspb.2001.1945

[49] Gandon S, Buckling A, Decaestecker E, Day T. Host–parasite coevolution and patterns of adaptation across time and space. Journal of Evolutionary Biology. 2008;**21**(6):1861-1866. DOI: 10.1111/j.1420-9101.2008.01598.x

[50] Marantos A, Mitarai N, Sneppen K. From kill the winner to eliminate the winner in open phage-bacteria systems. PLOS Computational Biology. 2022;**18**(8):e1010400. DOI: 10.1371/journal.pcbi.1010400

[51] Winter C, Bouvier T, Weinbauer MG, Thingstad TF. Trade-offs between competition and Defense specialists among unicellular planktonic organisms: The "killing the winner" hypothesis revisited. Microbiology and Molecular Biology Reviews. 2010;**74**(1):42-57. DOI: 10.1128/mmbr.00034-09

[52] de Jonge PA, Nobrega FL, Brouns SJJ, Dutilh BE. Molecular and evolutionary determinants of bacteriophage host range. Trends in Microbiology. 2019;**27**(1):51-63. DOI: 10.1016/j.tim.2018.08.006

[53] Liu R, Li Z, Han G, Cun S, Yang M, Liu X. Bacteriophage ecology in

biological wastewater treatment systems. Applied Microbiology and Biotechnology. 2021;**105**:5299-5307. DOI: 10.1007/s00253-021-11414-8/ Published

[54] Liu R, Qi R, Wang J, Zhang Y, Liu X, Rossetti S, et al. Phage-host associations in a full-scale activated sludge plant during sludge bulking. Applied Microbiology and Biotechnology. 2017;**101**:6495-6504. DOI: 10.1007/ s00253-017-8429-8

[55] Khot V, Strous M, Hawley AK. Computational approaches in viral ecology. Computational and Structural Biotechnology Journal. 2020;**18**:1605-1612. DOI: 10.1016/J.CSBJ.2020.06.019

[56] Runa V, Wenk J, Bengtsson S, Jones BV, Lanham AB. Bacteriophages in biological wastewater treatment systems: occurrence, characterization, and function. Frontiers in Microbiology. 2021;**12**:2708-2726. DOI: 10.3389/ fmicb.2021.730071

[57] Shapiro OH, Kushmaro A. Bacteriophage ecology in environmental biotechnology processes. Current Opinion in Biotechnology. 2011;**22**(3):449-455. DOI: 10.1016/J. COPBIO.2011.01.012